Laboratory Manual of
WATER AND WASTEWATER ANALYSIS

Today, water and waste water analysis has become increasingly important for may reasons. And for this reason, the objective has been to make available the information in single book. This book has been written to fulfill the need for a simple and concise manual on water and wastewater analysis. To make things easier for you there are flow diagrams of each parameter analysis.

It is hoped that this book would be useful for students, UG and PG level scientists, teachers, water analysts and for all those persons engaged in water and wastewater analysis.

Dr. D.R. Khanna, Ph.D, FIAES, FASEA, FNC, FZSI, FSES is Reader in Department of Zoology and Environmental Science, Gurukula Kangri Vishwavidyalaya, Haridwar. He is well known Limnologist and has published about 80 research papers and articles in national and international journals of repute. He has edited 6 books and authored 10 books. Dr. Khanna is Editor of Environment Conservation Journal and Associate editor of various journals. He was awarded ASEA excellence award, Nature conservators Impetus Gold medal and Bharat Jyoti award.

Dr. R. Bhutiani, FASEA, is Lecturer in the Department of Zoology and Environmental Science, Gurukula Kangri Vishwavidyalaya, Haridwar. Currently he is doing research in the field of limnological modeling. Dr. Bhutiani has authored/edited six books. He is associate editor of Environment Conservation Journal. He has published about 25 research papers and 5 articles in national and International Journals of repute. He has been awarded two junior young scientist awards for best research paper and appreciation award 2006.

Laboratory Manual of
WATER AND WASTEWATER ANALYSIS

Authors
D.R. Khanna
R. Bhutiani

2024
Daya Publishing House®
A Division of
Astral International Pvt. Ltd.
New Delhi – 110 002

ISBN: 9788170355281 (HB)

Published by : **Daya Publishing House®**
A Division of
Astral International Pvt. Ltd.
– ISO 9001:2015 Certified Company –
4736/23, Ansari Road, Darya Ganj
New Delhi-110 002
Ph. 011-43549197, 23278134
E-mail: info@astralint.com
Website: www.astralint.com

Printed at : **Replika Press Pvt Ltd**

Preface

Water is so common that we take it for granted. We probably think it is much like any other liquid, but it is not water is one of the most usual natural compound found on earth. Today, with the increase in population, water is exploiting at a faster rate as a result pollution load on water bodies is also increasing.

Therefore, to understand water as a necessity, as a reserve and a factor in the pollution problem, we must understand its characteristics, its role in the biosphere and its role in living things hence for their reasons water and wastewater analysis has become increasingly important.

Considerable literature is available regarding water and wastewater analysis but the need of a single comprehensive and concise laboratory manual for water and wastewater analysis has been felt by researchers and teachers. This book aims to provide the easiest and simplest way to analysis water and wastewater. Besides this the authors try to provide maximum information in a single book.

The book covers the various physico-chemical, biological and microbiological analysis of water and wastewater, routinely conducted in the laboratory. Flow diagram of analysis of parameters and data sheet for record are given at the end of each parametric details.

This book will be useful for the students of UG and PG level, teachers, water chemists, engineers, consultants, industrialists, academicians, scientists, researchers and for laboratories involved in the study of water and waste water analysis.

We urge our reader to send their feedback and suggestions for the improvement of this practical manual on any aspect including evaluation of the content, style of presentation, any errors, wrong formula, difficult terms topics to be included in next edition etc.

Your feedback will help us to improve the manual. All feedback can be sent to drkhanna2002@indiatimes.com or rbhutiani@gmail.com.

Lastly, it is pleasure to experts or sincere thanks to M/s Daya Publishing House, for their interests shown towards the publication of this book.

Haridwar, India

D.R. Khanna

R. Bhutiani

Contents

Chapter 1
Introduction

(a) Water and its Importance

As all of us know that about three-fourth of our earth is covered with water. The availability of this water in abundance has made the existence of life possible on earth for its creatures. It is one of the major natural gift for life on earth and is the only inorganic liquid and chemical compound which exists in all the three physical states *i.e.* solid, liquid, and vapour.

Nobel laureate A Szent-Gyorgy called Water "**The matrix of life**". Every living organisms need water for its existence as it is essential for the survival of any form of life. Human beings consume water for several purpose *viz.* drinking, bathing, irrigation, industrial process, transportation, waste disposal etc. Human beings inhabiting in tropical climate needs about two litres of water daily. Water is also the predominant constituent of living matter, seventy per cent of weight of human body is water.

Easy availability of water for drinking, farming, transportation etc. is an essential requirement of survival of an organism. It is estimated that about 1,011 million Km^3 of total water is present on earth, but out of this only 33,400 m^3 of water is available for drinking, agriculture, domestic and industrial purposes (Dara, 2001). The remaining water is available as Oceans (Salty water), polar ice-caps and glaciers.

Water is major requirement of any organism than any other resources, because it helps in many activities such as nourishment, sustainence of life, functioning of body organs, circulation of elements in the body etc. besides this it can easily be used as a flow resource and for removing waste. Also it is an important raw material of industries.

As it is clear from the above saying that no life could exist without water therefore for sustainence of life on earth water is must.

The easy availability and abundance of water on earth decreases the value of water as a resource as compared with other resources, water is used in tremendous quantities, because of its great abundance water is generally very inexpensive resource. With the increase in population our water resources are becoming polluted at an alarming rate which decrease the availability of fresh water. We are exploiting our water resources with great speed, which once seemed endlessly renewable when our population was limited.

Because the quality and quantity of water available at any particular time are highly variable, shortage of water have occurred and will probably continue to occur with increasing frequency.

(b) Today's Status of Water

Owing to increasing industrialization on one hand and exploding population on the other, the demand of water supply have been increasing tremendously. The total amount of water on earth is constant. It is fascinating fact that only 0.01% of the total water on earth is in the atmosphere. On a global scale, total water abundance is not the problem, the problem is water's availability in the right place at the right time in the right form.

Today with the unthoughtful use of water many countries located in arid and semiarid regions have started facing the problem of water crisis. As the total global population is increasing so do the aggregate human activities, would increase water requirements. On one hand due to increasing population water requirement is steadily increasing on the other hand, per-capita water available declines steadily since the total amount of fresh water available is limited.

The world's per-capita use of water in 1975 was about 700m^3 /year or 2000 litres/day (185,000gal/year) and the total human use of water was about 3850 Km^3/ year (about 1015 gal./year). The total average annual water yield (run-off) from Earth's rivers is approximately 47,000 Km^3 (1.2 × 10^{16} gal.), but its distribution is far from uniform (Botkin and Keller 2000).

Compared with other resources, water is used in tremendous quantities because of its great abundance water is generally very inexpensive resource.

Today the easy availability of fresh water is a major problem as 80% of our rivers are polluted due to industrial and domestic discharge. With increasing pollution problem availability of fresh water is decreasing at a greater rate. Today very rare or no river has been spared from industrial discharge this untreated industrial discharge increases the water pollutants such as high biochemical oxygen demand (BOD), acidity, alkalinity, hardness, heavy metals etc., presence of high values of these parameters makes the water unfit for human consumption. Domestic sewage also creates a major problem, as organic wastes from domestic water causes problem of eutrophication, so makes water unsuitable for consumption. Today most of the cities of India are facing water crisis and availability of fresh water for humans become a dream.

If this trend continues then in the next fifty years there may be possibility of a world war a **War on Water** and whole of our world would become desert where availability of water will not be an easy job.

So if we want to save our World from water crisis, then we would have to do something for it from now, the management should include controlled use, discharge after treatment and overall the human awareness without which nothing is possible.

(c) Sampling Procedures

Sampling procedures plays an important role in the analysis of physico- chemical parameters. The results of physico-chemical analysis of water depends upon the planning prior to take the sample. This should include number of samples, site selection, method of sample collection, parameters to be analysed, time to collect the sample and handling and preservation of sample. This ensures that the sample collected should have been identical to that of water body and have the same physico-chemical and bacteriological characteristics.

Therefore to avoid the change in characteristics of water sample much care should be taken while collection, handling and preserving the sample during transport (when sampling site is situated far away from laboratory) and to avoid accidental contamination, for collecting water samples always use water sampling bottle if not available use glass or plastic containers. To collect water sample always collect it from the bottom of water body (about 1/2 meter below the surface of water) to avoid changes in its characteristics.

Three types of sampling procedures are employed for analysis of water:

1. Spot or Grab Sample

Spot sample or grab sampling is the easiest type of sampling procedure in which the samples are collected from a particular place at a particular time. This type of sample is only valid when there is no change in the quality of water within short period of time if it happens samples must be collected at various time intervals and analyze. This type of sample collected at various time intervals may help in understanding the diurnal variations of several parameters and can be of greater value. The analysis of various number of grab samples should results in the accuracy, which satisfies in the requirements of sampling procedures.

2. Composite Sample

When a particular sampling site undergoes changes in its characteristics then the sample should be collected at various time intervals, the time interval must depend on the kind of analysis, it may vary from one, two, four, eight or twenty four hours. After collecting samples at various time intervals these may be combined to make a composite sample. The analysis of composite samples are useful for determining the average chemical characteristics.

3. Integrated Sample

When water samples are collected from different locations of the sampling stations and mixed thoroughly an integrated sample is prepared for analysis. For preparing integrated sample individual samples are collected from surface as well as from bottom of a shallow water body for a deep surface waterbody sampling at various depths is necessary. Both of the above mentioned techniques for preparing integrated sample helps in representing the same conditions of sample as found in natural water body.

Chapter 2
Monitoring

(a) Why the Monitoring of Water is Essential ?

Today with the increasing trend of population the requirement of water is also growing at a faster rate it is estimated that about 15-20 litres of water is consumed per person per day and after has been used by particular person, it becomes sewage or waste water. Sewage water includes all the water from a household that may be used for washing and toilet wastes while wastewater includes the industrial discharges.

Water that moves below the ground surface undergoes self purification by filteration method which removes many impurities of water including microorganisms due to self purification techniques water from deep wells is generally good in its quality. Water found in natural conditions is found seldom pure but the human pressure makes it unfit for any purpose such as drinking, bathing etc.

Today very rare or no river has been spared from contamination, the contamination may include untreated discharge of water from industries or from agricultural run-off. The discharge of industrial effluents into river water may result in increased values of various physico- chemical parameters which ultimately results in eutrophication of river and finally the river is termed as dead river (due to contamination of water upto a great extent). The use of pesticides in agricultural land also causes contamination of water, during rain, the pesticides mixes with river as run-off water of agricultural fields these pesticides when mixes with water, contaminates it and make it unfit for human consumption. Being non-biodegradable these pesticides results in biomagnification.

When water is obtained from uncontaminated reservoirs fed by clear mountain streams or from deep wells, it will require minimal treatment to make it fit for human consumption. But such type of reservoirs are very less in number, many cities, fulfill their water requirements from badly polluted sources such as rivers which are the dumping ground of many industrial and sewage wastes.

Therefore it is well understood from the above statement that why the monitoring of water is essential ? Historically most of our concern about water purity has been related to the transmission of disease. Therefore various tests have been developed to determine the safety of water. The major reason for monitoring the water quality is to see its pollution potential and to apply various methods to treat and make it fit for human consumption.

Today various attempts have been made by government to examine the physical, chemical, biological and microbiological characteristics of water and waste water. The water of river, lakes, ponds are monitored regularly to examine the pollution load by near by population. The industrial effluents are monitored so as to check the working efficiency of treatment plants.

Ultimately we must say that monitoring is essential to provide a safe and healthy life for human beings by providing them clean, disease free and nutrients rich water.

(b) Collection and Preservation of Samples

To analyse various parameters samples are collected under different conditions, the collection, handling and preservation of sample is most important factor related to sampling. The main objective of sampling is to collect sample, transport it carefully to the laboratory to analyse without effecting its quality, in other words we can say that the sample that we brought to the laboratory must represents the materials being sampled.

The sample must be handled in such a way that no significant change in its composition occurs before various tests are being made. To fulfill all the above requirements several precautions have been taken before sampling.

Let us start with the sampling equipments, various types of equipments are in use today, the detailed description of these equipments are given below but here a question arises, are these equipments are safe to collect sample ? The answer being No, it is not safe to collect sample before precleaning because mostly these equipments contains impurities which will dissolve with sample and ultimately results in error. So various precautions should be taken before handling these equipments such as to ensure the quality of containers and use containers that are clean and free of contaminants, for new containers rinse it with concentrated sulfuric acid (H_2SO_4) or concentrated Nitric acid (HNO_3) for old containers rinse it thrice with double distilled water.

Most common sampling equipments whose handling is easy are described below:

1. Mayer's Water Sampler (For Shallow Water System)

This is the simplest method for the collection of shallow water from required depth. The sampler consists of a glass bottle of two litre capacity the bottle is enclosed with a brass frame and weighted below to lower the bottle at any desire depth. The cork of the bottle is pulled up with the help of rope when filled with sample, the cork is lowered down to close the sampler. This sampler should be used upto a depth of 20 meter only.

Figure 2.1: Mayer's Water Sampler

2. Van Dorn Water Sampler

Van Dorn water sampler is very common type of sampler. It is made up of hollow cylinder of PVC with two cups attached with latex tube at both the ends. The sampler is lowered at desired depth, frees the cups with the help of rope. When cylinder gets filled with water the cup gets attached to close the cylinder due to elasticity of tube. The volume of the cylinder may vary from 0.5 to 5 litre or more.

3. The Nanseen Reversing Bottle Sampler

The nanseen reversing bottle sampler works more or less on the same principle as that of van dorn water sampler, it consists of reversing bottle which is made up of brass and interiorly coated with tin. It consists of two valves one on each end of reversing bottle. The sampler is lowered down with clamp of lower end with valve in open condition to collect water sample. It is held in this position by release mechanism but when lower end is released it strikes the release mechanism, which results the sampler falls over and turns to 180°ultimately shutting the valves of sampler. The bottle is brought up and the water is removed by opening the valves. The volume of sampler may vary from 0.5 to 1.5 litres.

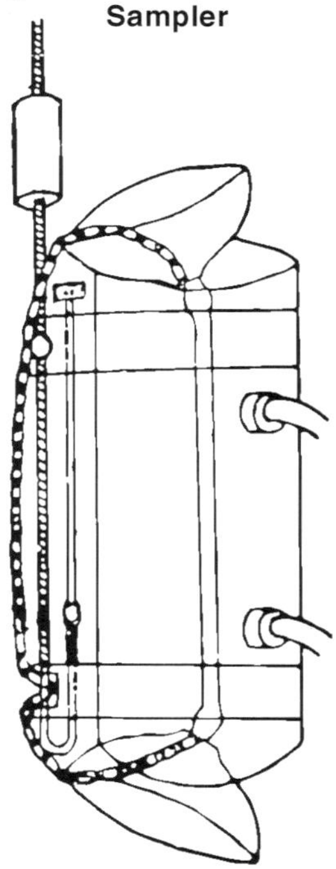

Figure 2.2: Van Dorn Water Sampler

4. Eckmen Dredge Sampler

It is very common and used for sampling soil sediments, it consists of steel box with two jaws joined together to the steel box. The sampler has fixed sectional area, the box is lowered down with jaws in open conditions when it reaches the target area messenger is sent down which closes the jaws collecting the desired sample after which it is pulled up to obtain sample.

It is advisable to collect 2&3 litres of sample for general analysis of physico-chemical parameters.

Safety measures should be taken during collection and preservation of sample because sometimes sampling body may effect our body organs, such as eyes, hands etc., therefore it must be important to know the characteristics of sample before collecting. Various safety measures such as use of hand gloves, mask, spectacles before sampling is must. Wash hand carefully after collecting sample, avoid taking food near sample. More details of precautions in the laboratory have been described in Chapter 9.

Note:

(*i*) Samples must be collected from 4-5 points from the same sampling site and mean value of these samples will minimize the error.

(*ii*) Sampling must be done between 7.00 A.M to 11.00 A.M.

(*iii*) Regular monitoring of the water bodies is necessary.

Chapter 3
Physical Analysis of Water

(a) Temperature (Surface/Non Depth Temperature)

Generally temperature is measured by good, callibrated, mercury thermometer. To measure surface temperature dip the mercury thermometer in the sample up to the level of mercury in capillary column and note the reading. It is advisable to use good quality thermometer *i.e.* well callibrated and accurate to minimize the error. For field operations always use thermometer having a metal case to prevent breakage. Regular callibration of thermometer is essential to avoid error in reading, it should be callibrated by another thermometer of known accuracy.

Depth temperature

Various types of devices are used to measure the depth temperature for limnological studies devices such as thermophone, thermister, reversing thermometer etc. are generally used, for accuracy and convience of environmentalist thermister is used, but due to its high cost it is very rare in use. The most common device for measuring depth temperature is reversing thermometer which is easily available and less in cost in comparison to thermistor or more important benefit of reversing thermometer is that it is often mounted on sample bottle so that sample collection and temperature of water may be taken simultaneously.

The correct reading from reversing thermometer for fluctuations between temperature at reversal and temperature at time of reading may be obtained by following equation:

$$T = T_1 + \Delta T$$

where,

$$\Delta T = \frac{(T_1 - t)(T_1 + V_o)}{(K - 100)} + I$$

where,

T: Corrected temperature of reversing thermometer.

T_1: Uncorrected reading of reversing thermometer.

t: Temperature at which thermometer is used

V_o: Volume of small bulb and of capillary upto 0°C graduation

K: Constant, depending upon the relative expansion of mercury and glass (Where K = 6100)

I: Callibration correction of thermometer

To avoid making calculations everytime it is convenient to obtain ΔT by preparing graph between T_1 and t in relation to ΔT.

Flow Diagram 3.1: Temperature Measurement

Collect suitable amount of sample in container

↓

Immediately dip mercury thermometer

↓

Note the reading in thermometer

Sheet No. 3.1: Data Sheet for Estimation of Temperature Measurement

Sl.No.	*Site/Source*	*Reading*						*Average*	*Standard Deviation*
		A	*B*	*C*	*D*	*E*	*F*		
1.									
2.									
3.									
4.									
5.									
6.									
7.									
8.									
9.									
10.									

±: Standard deviation.

(b) Colour

The colour in water may be from various contaminants such as suspended matter, natural metallic ions, humus, plankton, industrial wastes, collodial or suspended material. The colour in effluent of some industries is major problem colour is removed to make water suitable for some applications. It is generally advisable for some industries to discharge effluent after colour removal.

Two types of methods are generally used to determine colour:

(*i*) Visual Comparison Method

Principle

Visual comparison method is the easiest method to determine colour of water sample by comparing it with standards of known concentration. Platinum cobalt method is standard method to measure colour in standard solution one unit of colour is equivalent to 1mg platinum/litre.

As turbidity in water may change the values of colour may change therefore, it is advisable to remove the turbidity before calculating colour values, even slight turbidity may interfere with colour determination therefore for turbid water, sample must be centrifuged first for about 1 hour and supernatent of sample should be used to determine colour.

pH is an important factor while determining colour because colour value is totally pH dependent and is directly proportional to colour value *i.e.* as pH is increased colour value increases.Therefore specify pH at which colour is determined while determining colour value.

The visual comparison method is limited and is applicable to measure colour of potable water and of water in which colour is due to naturally occurring materials. This method is not applicable to highly coloured industrial waste effluents.

For analysis of colour, sample should be collected in clean glass bottles and analysed immediately because with increasing time colour should be changed biologically or physically.

Apparatus and Reagents Preparation

(*a*) Nessler's tubes, 50ml. capacity

(*b*) pH meter

(*c*) *Preparation of colour standards*: Dissolve 1.246 gms. of potassium chloroplatinate, K_2PtCl_6 and 1.0g crystallized cobaltous chloride, $CoCl_2.6H_2O$ in distill water add 100ml Concentrated HCl and finally make the volume to 1 litre in volumetric flask. This stock solution has 500 colour units.

Now prepare standards of 5, 10, 15, 20, 25, 30, 35, 40,,45, 50, 60 and 70 colour units by diluting 0.5, 1.0, 1.5, 2.0, 2.5, 3.0, 3.5, 4.0, 4.5, 5.0, 6.0 and 7.0 ml of stock coloured standard of 500 colour units with above quantity of distilled water. Put these standards in 50ml Nesslers tube and protect from evoparation and contamination.

Procedure

1. Take 50ml. of centrifuged sample in Nesslers tube.
2. If turbidity is not fully removed report colour as apparent colour.
3. Observe the sample by matching it with standard colour tubes by looking vertically both the tubes in bright sunlight or with white background.
4. If colour of the sample exceeds 70 units, dilute sample with distill water.
5. Measure the value of pH of all the samples.

Calculation

$$Colour\,Unit = \frac{A \times 50}{B}$$

where,

A: Estimated colour

B: ml. of sample taken for dilution

Flow Diagram 3.2: Colour Measurement (Method-1)

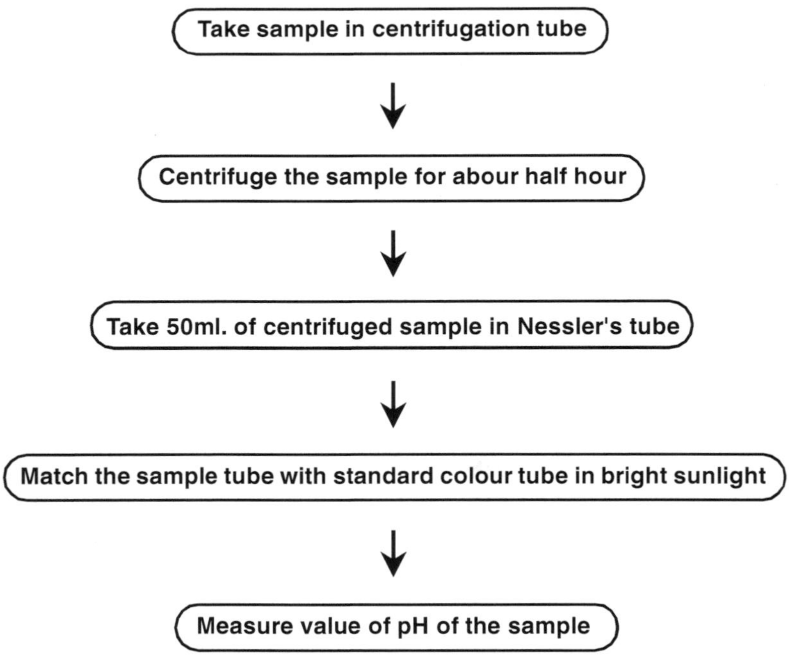

(*ii*) Spectrophotometric Method

Principle

The colour of a sample is expressed in terms that describe the sensation realized while viewing the sample. It may be expressed in terms llke luminance, dominant wavelength and by purity. These may be determined by viewing the sample by means of spectrophotometer. This method is applicable to all types of samples i.e industrial wastes, domestic wastes, potable water etc.

Apparatus

(*a*) Spectrophotometer

(*b*) Centrifuge

Procedure

(1) Centrifige Sample at 10,000 rpm for 30 minutes.

(2) Adjust the pH of sample to 7.6 using 0.2N NaOH or sulphuric acid.

(3) Now read the absorbance of supernatent at 465 nm under spectrophotometer.

(4) Run a blank using distill water, read absorbance at 465 nm.

Absorbance values are transfered in to colour unit (CU) using relationship:

$$C.U. = 500 \times \frac{A_2}{A_1}$$

A_1: Absorbance of 500 Cu platinum cobalt standard solution (0.132)

A_2: Absorbance of the effluent sample

Flow Diagram 3.3: Colour measurement [Method-2]

Take sample in centrifugation tubes

↓

Centrifuge the sample at 10,000 rpm for 30 minutes

↓

Adjust pH to 7.6 using 0.2N NaOH or H_2SO_4

↓

Read absorbance of supernatent at 465 nm under Spectrophotometer

Sheet No. 3.2: Data Sheet for Estimation of Colour

Sl.No.	Site/Source	Reading						Average	Standard Deviation
		A	B	C	D	E	F		
1.									
2.									
3.									
4.									
5.									
6.									
7.									
8.									
9.									
10.									

(c) Odour

Principle

The most common method to determine odour is "threshold odour number", which may be measured by diluting a particular ratio of sample with odour free water until the least perceptible odour is determined. This threshold method is applicable to all types of water ranging from odourless to highly odourless industrial or dometic waste water, only the highly odourless water are diluted before being presented to test observers. As the senstivity to odour varies from person to person, it is advisable to test sample from five to ten person. It should be noted that the observers are free from any kind of disease which will interfere in odour testing such as cold etc.

The threshold odour values varies with variation in temperature, most odours are detected at high temperature (60°C), for some odours a standard temperature of 40°C is beneficial.

Sample should be collected in clean glass bottles and are analysed as soon as possible, if not possible store the sample in refrigerator inside tight container to avoid entry of any other odour.

Apparatus

(*a*) Sample bottles

(*b*) Constant temperature bath or hot plate

(*c*) Odour flask-Glass stoppered 500ml capacity

(*d*) Measuing cylinders

(*e*) Pipets

(*f*) Thermometer

Procedure

(1) Firstly prepare the odour free water by filtering deionized, distilled or tap water through activated carbon.

(2) Determine approximate range of the threshold number by adding 200ml, 50ml, 12ml and 2.8ml of sample in separate 500ml, odours flask and finally make the volume to 200 ml by adding odour free water.

(3) Take a blank with odour free water in separate flask.

(4) Heat the flask on hot plate or water bath at desired temperature.

(5) Take the observation of maximum dilution at which odour is detected.

(6) Observe further dilutions on the basis of preliminary range of threshold odour number [T.O.N] and find out dilutions at which odour is detected including blank.

(7) Now observer must smell the flask from least concentration of sample towards higher concentration.

(8) Do not let testers know that which dilution is odorous and which is blank for getting accurate results.

(9) Record the observations according to the given table indicating whether odours is detected in each flask:

Ml of Sample diluted to 200 ml	12	0	17	25	0	35	50
Response	–	–	–	+	–	+	–

Threshold Odour Number Corresponding to Various Dilution

Sample Volume Diluted to 200 ml	*Threshold Odour Number*	*Sample Volume Diluted to 200 ml*	*Threshold Odour Number*
200	1	12	17
140	1.4	8.3	24
100	2.0	5.7	35
70	3.0	4.0	50
50	4.0	2.8	70
35	6.0	2.0	100
25	8.0	1.4	140
17	12.0	1.0	200

Source: Standard methods for examination of water and water, APHA, 20thed. 1998.

Calculation

$$\text{T.O.N} = \frac{A+B}{A}$$

where,

A: ml sample

B: ml odour free water

(A+B): 200ml

Flow Diagram 3.4: Odour measurement

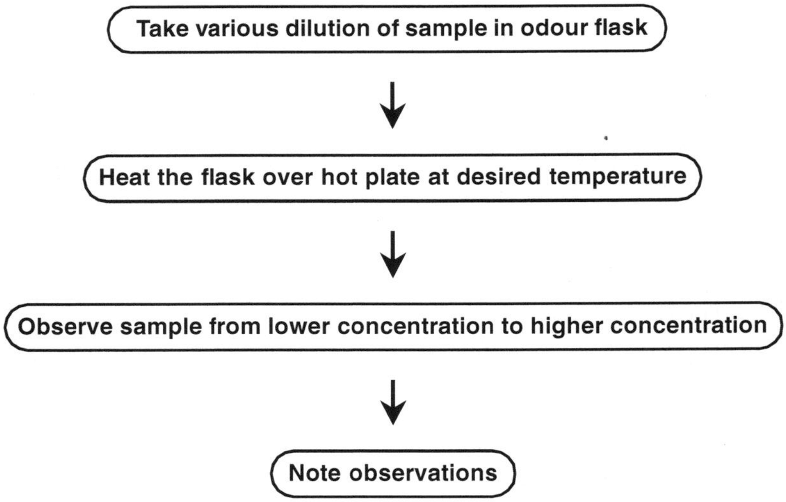

(d) Taste

Principle

Taste may be of several forms as bitter, salty, sour or sweet it will result from chemical sense upon contact with receptor organ, water sample taken into mouth produces flavour therefore it can be analysed only by tasting a sample. However all the water samples could not be tasted, as some of them consists of harmful chemicals or pathogens.

Apparatus

(*a*) Constant temperature bath or hot plate

(*b*) Beaker [50 ml]

(*c*) Measuring cylinder

(*d*) Pipettes

Procedure

(1) Prepare the dilutions as prepared in odour analysis.

(2) Heat them to desired temperature.

(3) Give dilutions to observers in 50ml beakers, including blank.

(4) Let the observer taste the sample in sequence from lower concentration to higher concentration.

(5) Observer should take the suitable volume of sample in to his mouth and gives result without swallowing the sample.

(6) The result should be compared with blank.

Calculation

$$\text{F.T.N (Flavour threshold number)} = \frac{A + B}{A}$$

where,

A: ml of sample volume

B: Dilution water (ml)

Flow Diagram 3.5: Taste measurement

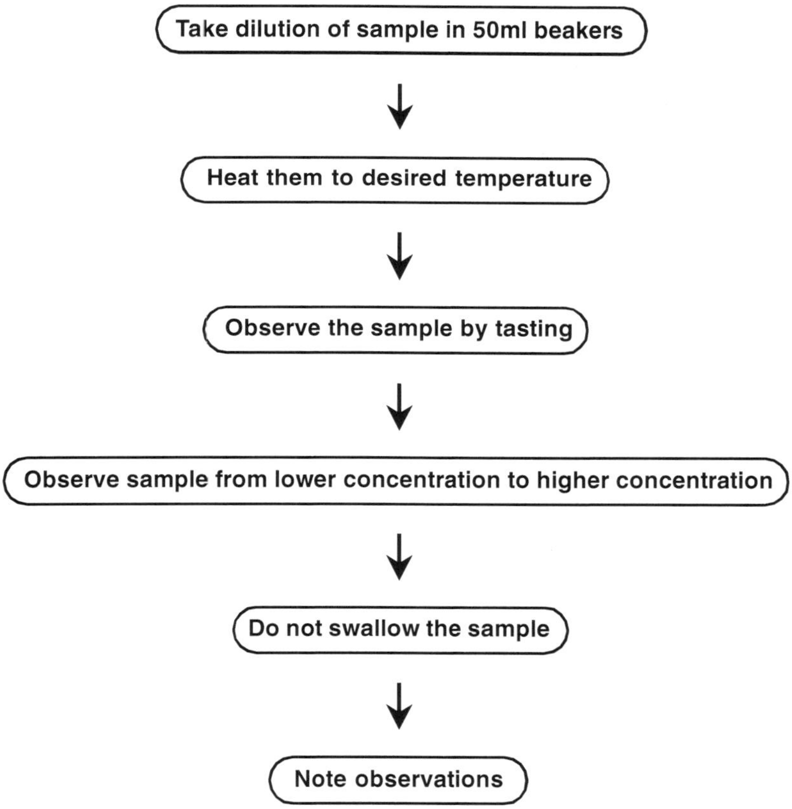

Sheet No. 3.3: Data Sheet for Estimation of Taste

Sl.No.	Site/Source	Reading						Average	Standard Deviation
		A	B	C	D	E	F		
1.									
2.									
3.									
4.									
5.									
6.									
7.									
8.									
9.									
10.									

(e) Turbidity

(*i*) Nepthelometer Method

Principle

In this method, when light is passed through a turbid sample, the particles present in sample scatter some of the light and the light scattered is directly proportional to the turbidity *i.e.* higher the turbidity, higher will be the light scattered, the amount of light scattered by the sample is compared with the intensity of light scattered by a standard suspension.

Turbidity can be determined for any kind of water sample, but it should be debris and rapidly settlable particles.

Apparatus and Reagents Preparation

(a) Nephelometer

(b) Sample tubes: Should be free from any scratch

(c) *Stock turbidity suspension:*

(*i*) Dissolve 1.0 gm of hydrazine sulphate $(NH_2)_2 . H_2SO_4$ in distilled water to prepare 100ml. of solution using volumetric flask.

(*ii*) Dissolve 10.0 gm hexamethyline tetramine $(CH_2)_6N_4$, in distilled water to prepare100ml. of solution in volumetric flask.

Mix 5ml. of solution (*i*) and 5ml. of solution (*ii*) in volumetric flask. Allow to stand for one day at 25±3°C dilute it to 100ml, this is 400 NTU (Nephelometric turbidity unit) standard suspension. This solution is stable upto one year if properly stored.

(d) *Standard turbidity suspension:* To prepare turbidity suspension dilute 10ml. of stock solution to 100ml. the solution is of 40 NTU and should be prepared freshly and discard it after use.

Procedure

1. Read manual and follow the instructions given in it to handle nephelometer.
2. Set the instrument at 100 using 40 NTU standard turbidity suspension. Every division on the scale will be equal to 0.4 NTU turbidity.
3. Remove air bubble before measurement.
4. Put the sample filled tube inside nephelometer and note the reading.
5. If sample is more turbid, dilute it and take reading.

Calculation

Turbidity (NTU) = Nephelometer reading $\times$ 0.4 $\times$ dilution factor

Flow Diagram 3.6: Turbidity Measurement (Method-1)

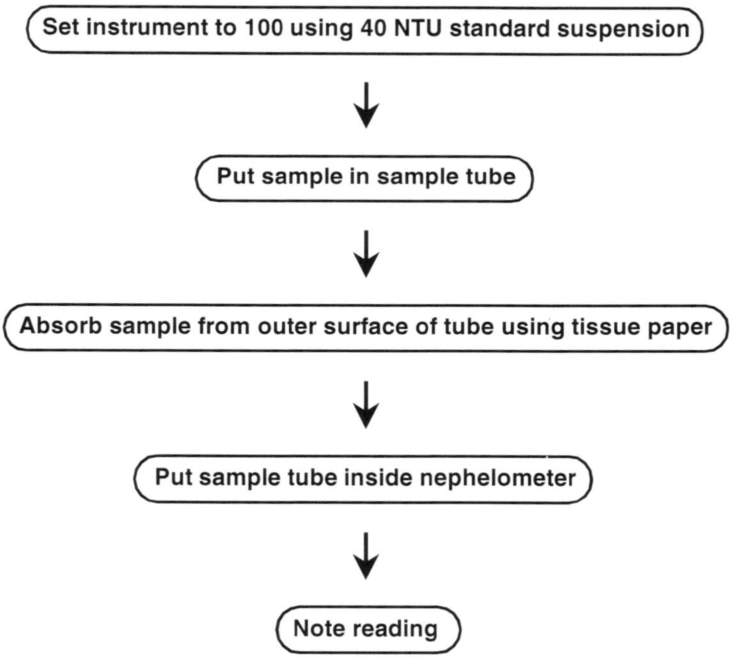

(*ii*) Jackson Turbidity Meter

It is most common instrument to measure turbidity. The instrument consists of three parts callibrated glass tube, a holder and a candle. The glass tube and the candle are supported in a vertical position.

Pour sample inside the glass tube until image of flame of candle disappears. Note the reading from glass tube in centimeters and convert it to J.T.U. (Jackson turbidity unit) using following Table 3.1.

Table 3.1: Conversion Table for Turbidity (Centimeters into JTU)

Cm	*Turbidity*	*Cm*	*Turbidity*	*Cm*	*Turbidity*	*Cm*	*Turbidity*
2.3	1000	6.3	350	10.3	210	25.1	85
2.6	900	6.4	340	10.8	200	26.7	80
2.9	800	6.6	330	11.4	190	28.1	75
3.2	700	6.8	320	12.5	180	29.8	70
3.5	650	7.0	310	12.7	170	31.8	65
3.8	600	7.3	300	13.5	160	34.1	60
4.1	550	7.5	290	14.4	150	36.7	55
4.5	500	7.8	280	15.4	140	39.8	50
4.9	450	8.1	270	16.6	130	43.5	45
5.5	400	8.4	260	18.0	120	48.1	40
5.6	390	8.7	250	19.6	110	54.0	35
5.8	380	9.1	240	21.5	100	61.8	30
5.9	370	9.5	230	22.6	95	72.9	25
6.1	360	9.9	220	23.8	90		

Source: Water and waste water testing (Mathur, 1982).

Figure 3.1: Jackson's Turbidity Meter

Flow Diagram 3.7: Turbidity Measurement (Method-2)

Shake the sample vigorously

↓

Transfer slowly the sample into glass tube of Jackson's Candle turbidimeter

↓

Add sample slowly until the flame disappears

↓

Note the reading in cms.

↓

Express reading using conversion table

Sheet No. 3.4: Data Sheet for Estimation of Turbidity

Sl.No.	*Site/Source*	*Reading*						*Average*	*Standard Deviation*
		A	*B*	*C*	*D*	*E*	*F*		
1.									
2.									
3.									
4.									
5.									
6.									
7.									
8.									
9.									
10.									

(f) Conductivity

Principle

Water usually consists of various types of ions, these ions helps in passing the electric current through water. Hence, conductivity may be defined as the measure of ability of aqueous medium to carry on electric current. Conductivity is totally dependent upon concentration of ions.

Conductivity is generally measured by conductivity meter, which consists of conductance cells having electrodes of platinum, these electrodes are placed parallely at a fixed distance.

As the ionization of solutes are totally dependent on temperature, therefore all the results are measured at 25° C. The unit of conductivity is siemens/cm or µmho/cm.

Apparatus and Reagent Preparation

(*a*) Conductivity meter

(*b*) Thermometer

(*c*) Conductivity cells

(*d*) *Standard potassium chloride solution (0.01M)*: Dissolve 745.6 mg anhydrous KCl solution in distill water and make volume to 1000ml. in volumetric flask and store it at 25 °C in CO_2 free atmosphere. This solution has conductivity of 1412 µmhos/cm at 25° C.

Procedure

1. Rinse conductivity cell with 0.01M KCl solution.
2. Adjust the temperature to 25 °C.
3. Measure the resistance of KCl solution.
4. Compute cell constant with following formula:-

   Cell constant

   $C, (cm^{-1}) = (0.001412)\,(R_{KCl})\,[1 + 0.0191\,(t - 25)]$

where,

R_{KCl}: Measured resistance of KCl

t: temperature (°C)

(5) Now again rinse the conductivity cell with sample water.

(6) Adjust the temperature to 25 °C.

(7) Measure the resistance of the sample.

Calculation

$$Conductivity\,(K)\,\mu mhos = \frac{1{,}000{,}000 \times C}{R_s[1 + 0.019]\,(t - 25)}$$

where,

K: Conductivity
C: Cell constant
R_s: Measured resistance of sample
t: temperature (°C)

Flow Diagram 3.8: Conductivity Measurement

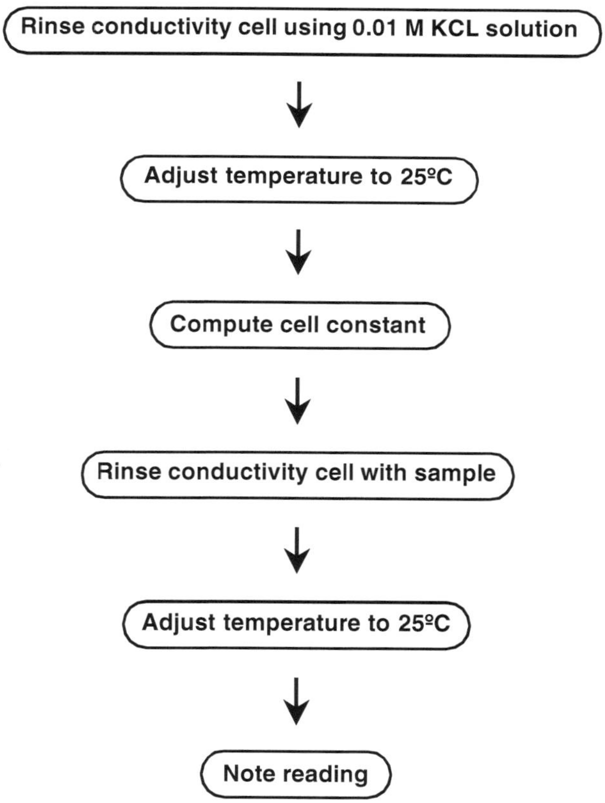

Sheet 3.5: Data Sheet for Estimation of Conductivity

Sl.No.	*Site/Source*	*Reading*						*Average*	*Standard Deviation*
		A	*B*	*C*	*D*	*E*	*F*		
1.									
2.									
3.									
4.									
5.									

Contd...

Sheet 3.5–Contd...

Sl.No.	Site/Source	Reading						Average	Standard Deviation
		A	B	C	D	E	F		
6.									
7.									
8.									
9.									
10.									

(g) Total Solids

Principle

The solids represents a portion of water sample that are not lost even after evaporation. Therefore we may find total solids as residue left after evaporating the unfiltered known volume of sample. The increase in weight over that of empty dish represents weight of total solids.

Apparatus

(*a*) Evaporating dish
(*b*) Oven
(*c*) Electronic balance
(*d*) Measuring cylinder

Procedure

(1) Take an evaporating dish and weight it.
(2) Now take 50ml. of unfiltered sample into it.
(3) Evaporate the sample at 100-103°C in an oven.
(4) Cool it and weight again.

Calculation

$$Total\ Solids\ (mg/l) = \frac{(A-B)\times 1000}{Volume\ of\ Sample\ (ml)}$$

where,

A: Final weight of dish (with sample)
B: Previous weight of dish (without sample)

Flow Diagram 3.9: Total Solids Measurement

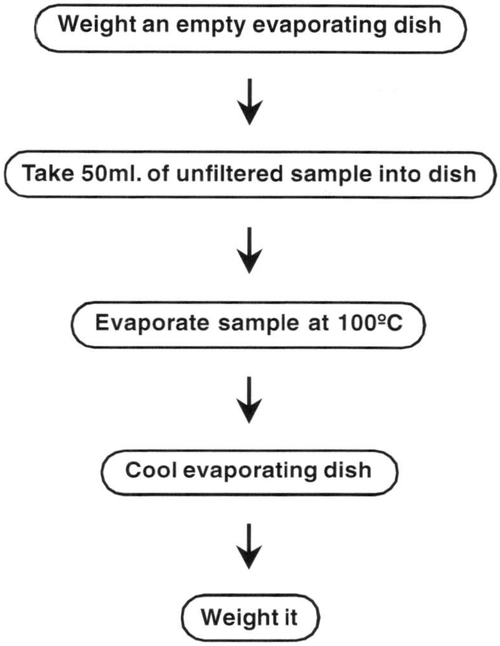

Sheet No. 3.6: Data Sheet for Estimation of Total Solids

Sl.No.	*Site/Source*	*Reading*						*Average*	*Standard Deviation*
		A	*B*	*C*	*D*	*E*	*F*		
1.									
2.									
3.									
4.									
5.									
6.									
7.									
8.									
9.									
10.									

(h) Total Dissolved Solids

Principle

Total solids in any sample can be represented by dissolved and particulate organic and inorganic matter and may be estimated by evaporating the filtered sample through standard filter and weighing the residue left.

Apparatus

(*a*) Evaporating dish
(*b*) Oven
(*c*) Electronic balance
(*d*) Measuring cylinder
(*e*) Filter paper (Standard)

Procedure

(1) Take an evaporating dish and weight it.
(2) Now take 50ml. of filtered sample in it.
(3) Evaporate the sample in an oven at 100-103°C or on water bath.
(4) Cool the evaporating dish and weight it.

Calculation

$$T.D.S.(mg/l) = \frac{(A-B)\times 1000}{Volume\ of\ Sample\ (ml)}$$

where,

A: Final weight of dish (with sample)
B: Initial weight of dish (without sample)

Flow Diagram 3.10: Total Dissolved Solids Measurement

Weight on empty evaporating dish

↓

Take 50ml. of filtered sample in to dish

↓

Evaporate sample at 100°C

↓

↓

Cool evaporating dish

↓

Weight it

Sheet No. 3.7: Data Sheet for Estimation of Total Dissolved Solids

Sl.No.	*Site/Source*	*Reading*						*Average*	*Standard Deviation*
		A	*B*	*C*	*D*	*E*	*F*		
1.									
2.									
3.									
4.									
5.									
6.									
7.									
8.									
9.									
10.									

(i) Total Suspended Solids (T.S.S)

Principle

Filter sample through the preweighted filter paper then dry and weight the filter paper, the increase in weight of filter paper represents total suspended solids.

Apparatus

(*a*) Measuring cylinder
(*b*) Filter paper
(*c*) Oven
(*d*) Beaker

Procedure

(1) Weight the filter paper of standard size.
(2) Now filter 50ml. of sample through preweighted filter paper.

(3) Dry filter paper in an oven at 90°C for 1 hour.

(4) Weight filter paper after drying.

Calculation

$$T.S.S.(mg/l) = \frac{(X - Z) \times 100}{Volume\ of\ \ Sample\,(ml)}$$

where,

X: Final weight of filter paper

Z: Previous weight of filter paper

Flow Diagram 3.11: Total Suspended Solids Measurement

Weight the filter paper of standard size

↓

Filter 50ml. of sample through it

↓

Dry filter paper in an oven at 90°C for 1 hour

↓

Weight dry filter paper

Sheet No. 3.8: Data Sheet for Estimation of Total Suspended Solids

Sl.No.	*Site/Source*	*Reading*						*Average*	*Standard Deviation*
		A	*B*	*C*	*D*	*E*	*F*		
1.									
2.									
3.									
4.									
5.									
6.									
7.									
8.									
9.									
10.									

(j) Velocity

Principle

The particular distance covered by a preweighted drifting float in a particular time represents the velocity of the running water. The float may be of any size or shape, are previously weighted so as to get them submerged and have no wind effect.

Apparatus

(*a*) Inch tape

(*b*) Float tables (flower)

(*c*) Stop watch

Procedure

(*a*) Measure the distance at which velocity has to be made.

(*b*) Install poles at both the ends.

(*c*) Allow the weighted float to drift from one pole to another.

(*d*) Note the time to cover distance from one pole to another (In minutes)

(*e*) Repeat the process several times to obtain average value.

Calculation

$$Velocity(m/s) = \frac{D}{T}$$

where,

D: Distance covered by float in meters

T: Time taken by float to cover distance from one pole to another (in minimum)

Flow Diagram 3.12: Velocity Measurement

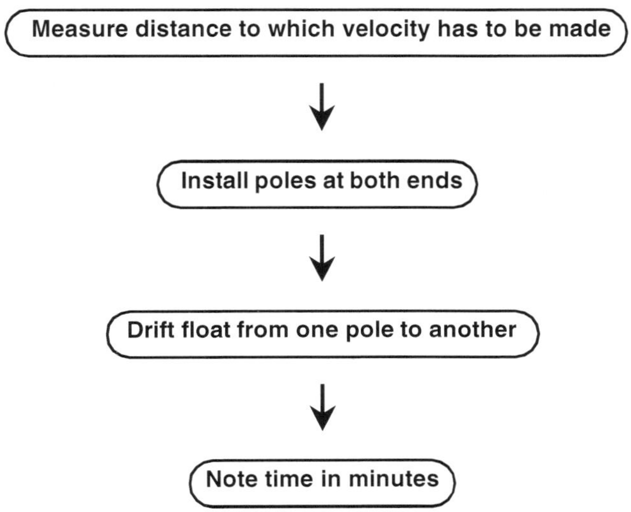

Sheet 3.9: Data Sheet for Estimation of Velocity

Sl.No.	Site/Source	Reading (in minutes)						Average	Standard Deviation
		A	B	C	D	E	F		
1.									
2.									
3.									
4.									
5.									
6.									
7.									
8.									
9.									
10.									

Chapter 4
Chemical Analysis of Water

(a) pH (Potentia Hydrogenii)

pH is the scale of acidity and alkalinity which defines the medium of the sample. pH can be measured either by electrometric method or by using pH strips or various indicators.However for best results always use electrometric methods.

Apparatus

1. pH meter/pH strips
2. Glass electrode
3. Temperature compensating device

Procedure

(*a*) Always follow the manufacturers instruction manual for using pH meter.

(*b*) Calibrate the instrument before each measurement. If the pH values vary widely, calibrate the instrument every time using suitable buffers.

(*c*) Rinse the electrode with distill water, dry it with soft tissue paper.

(*d*) Dip the electrode on sample water and note down the reading.

Note: Always dip the electrodes in distill water upto 2/3rd of its length before and after using it, otherwise it will cause damage to it.

Flow Diagram 4.1: pH Measurement

Calibrate pH meter following manufacturers manual

↓

Wash electrode with distill water

↓

Dry electrode using tissue paper

↓

Dip half electrode in sample water

↓

Note the reading

Sheet 4.1: Data Sheet for Estimation of pH

Sl.No.	*Site/Source*	*Reading*						*Average*	*Standard Deviation*
		A	*B*	*C*	*D*	*E*	*F*		
1.									
2.									
3.									
4.									
5.									
6.									
7.									
8.									
9.									
10.									

(b) Dissolved Oxygen (DO)

Most commonly used method for calculating DO in water body is winkler's Iodometric method.

Principle

The iodometric test for DO is based upon addition of manganous sulphate immediately followed by strong alkali to the water sample in measured quantity, it will form a white precipitate which in presence of oxygen gets oxidised to brown colour compound, finally this oxidised manganous gets reverts to iodine by iodide ions this iodine will be equivalent to the original DO content. The iodine is then titrated against thiosulphate.

Apparatus

(*a*) DO bottles (300 ml capacity)

(*b*) Glass bottles should be preferred

(*c*) Conical flask

Reagents

(A) Sodium Thiosulphate, 0.025 N

Dissolve 24.82gms of $Na_2S_2O_3.5H_2O$ in boiled distilled water now add 0.4 gms solid NaOH to it finally dilute the solution to one litre, this is stock solution dilute 250 ml. of this stock solution to one litre, it is 0.025N Sodium thiosulphate solution.

(B) Manganous Sulphate Solution

Dissolve 100 gms. of $MnSO_4.4H_2O$ in 100ml. of boiled distill water, filter the solution immediately after dissolving.

(C) Alkaline Potassium Iodide Solution

Dissolve 100 gms. of potassium hydroxide (KOH) and 50 gms. of potassium iodide (KI) in 200ml. of boiled distilled water.

(D) Starch Solution

Dissolve 1gm of starch to 100ml. of warm distilled water add few drops of formal dehyde solution.

*Instead of aqueous solution, soluble starch powder can also be used.

(E) Sulphuric Acid

Concentrated.

Procedure

1. Fill completely the BOD bottle (300 ml.) avoiding any kind of air bubble inside it and close the mouth of bottle inside water body.
2. Now pour 2ml. of $MnSO_4$ and alkaline potassium iodide to the sample through the walls of bottle. A precipitate will appear after few seconds.

3. Mix the contents thoroughly, now place the bottle for sometime so that the precipitate will settle down **(Here oxygen get fixed).**
4. Add 2ml. of concentrated H_2SO_4 and shake the bottle to dissolve the precipitate.
5. Take either whole or part of sample which ever necessary to a conical flask for titration.
6. Titrate the sample against sodium thiosulphate using starch as an indicator.
7. At end point, initial dark blue colour changes to colourless.

Calculation

When whole contents have been titrated:

$$DO(mg/l) = \frac{(ml \times N\ of\ titrant) \times 1000 \times 8}{V_1 - V}$$

When only part of content have been titrated

$$DO(mg/l) = \frac{(ml \times N\ of\ titrant) \times 1000 \times 8}{V_2\left(\frac{V_1 - V}{V_1}\right)}$$

where,

V_1: Total volume of sample in bottle (300 ml.)
V_2: Volume of sample titrated
V: Volume of $MnSO_4$ and KI added (4ml).
N: Normality

Flow Diagram 4.2: DO Measurement

Take sample in 300ml capacity BOD bottles

↓

Pour 2ml. of each $MnSO_4$ and alkaline KI solution (reagents should be poured at the bottom from the wall)

↓

Mix Sample thoroughly

↓

↓

A precipitate will appear

↓

Keep the bottle to settle the precipitate

↓

Add 1-2ml. of concentrated H_2SO_4 and shake well to dissolve the precipitate

↓

Take 50-100 ml. of sample in a conical flask

↓

Titrate with Sodium thiosulphate solution using starch as an indicator

↓

Dark blue colour changes to colourless at end point

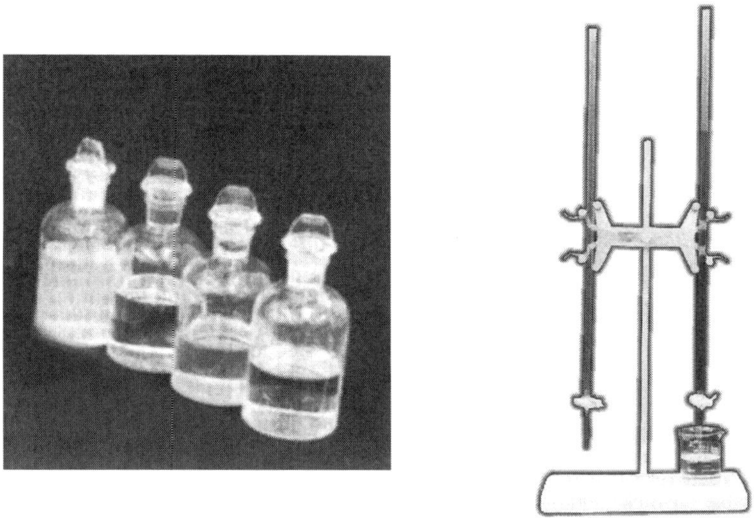

Figure 4.1: Titration Apparatus and BOD Bottles for DO and BOD Analysis

Sheet 4.2: Data Sheet for Estimation of Dissolved Oxygen

Sl.No.	Site/Source	Reading						Average	Standard Deviation
		A	B	C	D	E	F		
1.									
2.									
3.									
4.									
5.									
6.									
7.									
8.									
9.									
10.									

(c) Biochemical Oxygen Demand (BOD)

The biochemical oxygen demand (BOD) determination is an empirical test in which standardized laboratory procedures are used to determine the relative oxygen requirement of wastewater effluents, and polluted waters.

The test measure the molecular oxygen utilized during a specified incubation period for the biochemical degradation of organic matter (carbonaceous demand) and the oxygen used to oxidize inorganic material such as sulfides and ferrous ions. In other words biological oxygen demand, (BOD) represents the amount of oxygen required for the microbial decomposition of the organic matter in the water.

The BOD procedure, which is used extensively in monitoring water quality and biodegradation of waste material, is designed to determine how much oxygen microorganisms consume during oxidation of the organic matter in the sample. For industrial effluent or highly polluted sample dilute it with distill water.

Incubate BOD sample for 5 days in BOD incubator at 20°C and follow same steps as in measurement of dissolved oxygen.

Calculation

$$\text{BOD, mg/l} = (D_0 - D_5) \times \text{dilution factor}$$

Where,

D_0: Initial DO in the sample

D_5: DO after five days incubation

(d) Chemical Oxygen Demand (COD)

Principle

The amount of Oxygen required to completely oxidize the organic matter present in a sample is called Chemical oxygen demand (COD). All the organic compounds in water can be oxidised to CO_2 and H_2O in presence of strong oxidizing agent as potassium dichromate under acidic conditions.

For samples COD of > 900 mg/l, use smaller portion diluted to 50.0 ml.

Apparatus

1. COD assembly.
2. Measuring cylinder
3. Conical flask
4. Beaker
5. Dropper
6. Burette
7. Pipette

Reagents

A. Potassium Dichromate Solution 0.25 N

Dissolve 12.25g (previously oven dried) $K_2Cr_2O_7$ is distilled water and make volume to 1 litre.

B. Potassium Dichromate Solution 0.025 N

Dilute 0.25 N $K_2Cr_2O_7$ to 10 times to make 0.025 N $K_2Cr_2O_7$

C Ferrous Ammonium Sulphate 0.1 N

Dissolve 39.2 g of Fe $(NH_4)_2$ $(SO_4)_2$. $6H_2O$ in distilled water and add 20ml. of concentrated H_2SO_4 finally make volume to 1 liter using distilled water. Standardize this solution daily.

D. Ferrous Ammonium Sulphate 0.01 N

Dilute 0.1N $Fe(NH_4)_2$. $(SO_4)_2$. $6H_2O$ 10 times (100-1000ml.)

E. Ferroin Indicator

Dissolve 1.485g of 1,10 phenolphthaline monohydrate and 695 mg $FeSO_4.7H_2O$ in distilled water and dilute it to 100ml.

F. Sulphuric Acid

Concentrated.

G. Mercuric Sulphate

Solid.

H. Silver Sulphate

Solid.

Procedure

1. Take 50ml. sample in a 500ml. refluxing flask (for samples with COD more than 900mg/l take small amount of sample diluted to 50 ml.
2. Add 1gm of mercuric sulphate, several glass beads and 1gm silver sulphate.
3. Very slowly add 5ml. H_2SO_4 to dissolve $HgSO_4$.
4. Add 10ml. of $K_2Cr_2O_7$
5. Attach flask to condensor and turn on cooling water.
6. Add remaining H_2SO_4 (70ml.) through open end of condensor.
7. Cover open end of condensor with a beaker to prevent foreign material from entering refluxing mixture.
8. Reflux for 2 hrs.
9. Cool and wash down condensor with distilled water.
10. Remove the flask, cool it and dilute mixture to about twice its volume with distilled water.
11. Titrate excess $K_2Cr_2O_7$ with FAS using 2-3 days of ferroin indicator.
12. At the end point colours changes to reddish brown.
13. Run a blank with distilled water using same amount of chemicals.

Calculation

$$\text{COD}\,(mg/l) = \frac{(A - B) \times N \text{ of } FAS \times 8000}{ml \text{ of sample taken}}$$

where,

A: ml. of FAS used for blank

B: ml. of FAS used for sample

Flow Diagram 4.3: COD Measurement

Place 50 ml. samples in a 500 ml refluxing flask

↓

Add 1 g $HgSO_4$

↓

Several glass beads

↓

↓

**Very slowly add 5.0 ml sulphuric acid reagent mix to dissolve $HgSO_4$
Cool while mixing to avoid possible loss of volatile materials**

↓

Add 25 ml 0.0417 M $K_2Cr_2O_7$ Solution & mix

↓

Attach flask to condenser and turn on cooling water

↓

Add remaining Sulphuric acid reagent (70ml) through open end of condenser

↓

Cover open end of condenser

↓

Reflux for 2 hrs.

↓

Cool & wash down condenser with distilled water

↓

Disconnect reflux condenser and dilute mixture to about twice its volume with distilled water

↓

Cool to room temperature

↓

Titrate excess $K_2Cr_2O_7$ with FAS using 0.10 to 0.15ml Feroin indicator (2-3 drops)

↓

End point reddish brown

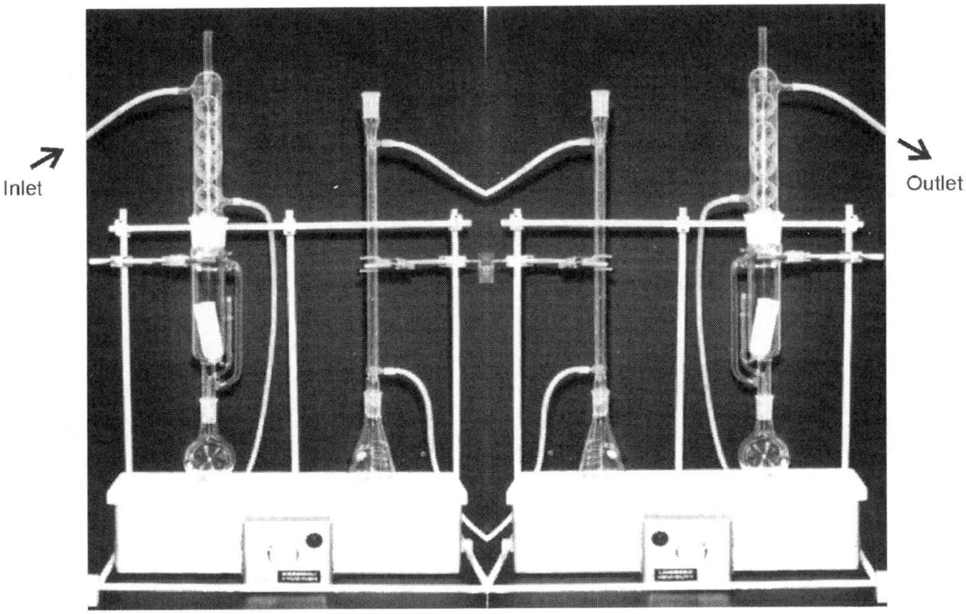

Figure 4.2: Diagram Showing COD Assembly

Sheet 4.3: Data Sheet for Estimation of Chemical Oxygen Demand

Sl.No.	Site/Source	Reading						Average	Standard Deviation
		A	B	C	D	E	F		
1.									
2.									
3.									
4.									
5.									
6.									
7.									
8.									
9.									
10.									

(e) Free Carbon Dioxide

Principle

Free carbondioxide can easily be determined by reacting the sample with Sodium hydroxide which ultimately forms Sodium bicarbonate.

Apparatus

(*a*) Conical Flask

(*b*) Burette

(*c*) Pipette

(*d*) Dropper

Reagents

(A) Sodium Hydroxide 0.05 N

Dissolve 40 gm of NaOH in boiled distilled water this is 1.0 N NaOH stock solution. Now take 50ml. stock solution and dilute it to 1 litre to prepare 0.05N NaOH solution.

(B) Phenolphthalein Indicator

Dissolve 0.5 gm of phenophthalein in 50ml. ethanol followed by 50ml. distill water further mix 0.05N NaOH dropwise until solution becomes faintly pink.

Procedure

1. Take 100ml. of sample in a conical flask.
2. Add 2-3 drops of phenolphthalein indicator.
3. If colour of sample turns pink free CO_2 is absent.
4. If sample remains colourless titrate with 0.05N NaOH.
5. At end point colourless solution turns pink.

Calculation

$$Free\ CO2\ mg/l = \frac{(ml \times N\ of\ titrant) \times 1000 \times 44}{ml\ of\ sample\ taken}$$

Flow Diagram 4.4: Carbon Dioxide Measurement

Take 100 ml of Sample in a Conical flask

↓

Add few drops of phenolphthalein indicator

↓

Colour turns pink, free CO_2 is absent

↓

↓

If Sample remains colourless

↓

Titrate it against 0.05 N NaOH

↓

Colourless solution turns to pink at end point

Sheet 4.4: Data Sheet for Estimation of Free Carbon Dioxide

Sl.No.	Site/Source	Reading						Average	Standard Deviation
		A	B	C	D	E	F		
1.									
2.									
3.									
4.									
5.									
6.									
7.									
8.									
9.									
10.									

(f) Acidity

Principle

The capacity to react with a strong base to a fixed pH is called acidity of water. Acidity in water result in change of its quality which may be due to strong or weak acids and also due to presence of some hydrolyzing salts. In sample of industrial wastes the acidity is due to presence of hydrolyzable metal ions whereas in freshwater the acidity is due to presence of free CO_2.

Apparatus

1. Conical flask

2. Burette
3. Pipette
4. Dropper

Reagents

(A) Sodium Hydroxide 0.05 N

See in free CO_2.

(B) Methyl Orange Indicator

Dissolve 0.5 g of methyl orange indicator and dilute it in 100ml distill H_2O. This is 0.05% methyl orange indicator.

(C) Phenolphthalein Indicator

See is free CO_2.

Procedure

(*a*) Take 100ml. of sample in a conical flask.
(*b*) Add 2-3 drops of methyl orange indicator.
(*c*) If solution turns yellow methyl orange acidity is absent.
(*d*) If solution turns pink, titrate it with 0.05N NaOH until pink colour changes to yellow.
(*e*) Add 2-3 drops of phenolphthalein indicator.
(*f*) Titrate further with 0.05 N NaOH.
(*g*) At end point yellow colour changes to pink.

Calculation

$$Methylorange\ \ acidity\,(mg/l) = \frac{(V \times N\ of\ NaOH) \times 1000 \times 50}{ml\ of\ sample\ taken}$$

$$Phenolphthalein\ acidity(mg/l) = \frac{V1 \times (N\ of\ NaOH) \times 1000 \times 50}{ml\ of\ sample\ taken}$$

$$Total\ acidity\,(mg/l) = \frac{(V + V1) \times N\ of\ NaOH \times 1000 \times 50}{ml\ of\ sample\ taken}$$

where,

V: Vol. of NaOH used with methyl orange
V_1: Vol. of NaOH used with phenolphthalein

Flow Diagram 4.5: Acidity Measurement

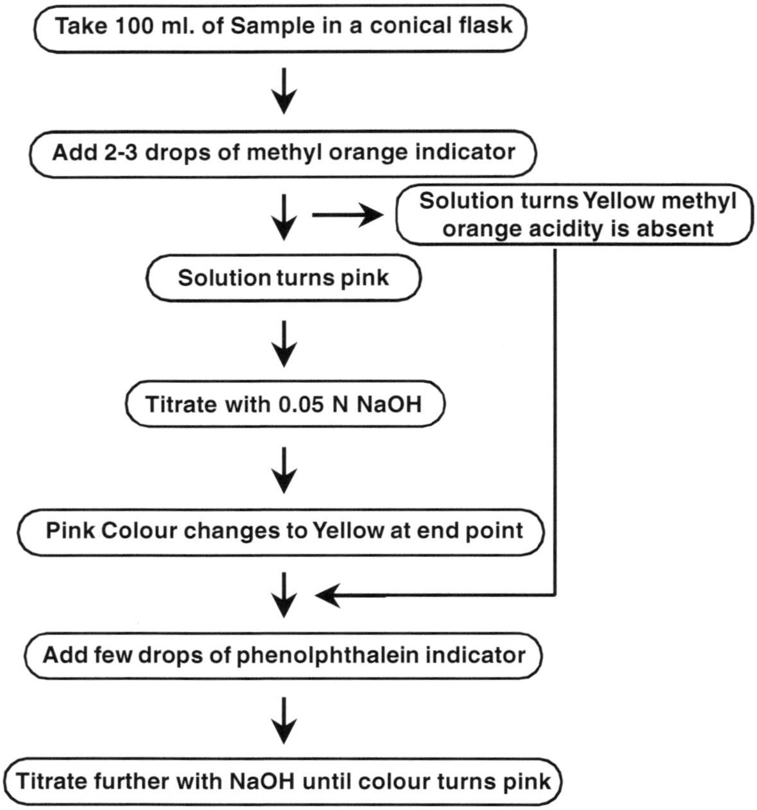

Sheet 4.5: Data Sheet for Estimation of Acidity

Sl.No.	Site/Source	Reading						Average	Standard Deviation
		A	B	C	D	E	F		
1.									
2.									
3.									
4.									
5.									
6.									
7.									
8.									
9.									
10.									

(g) Alkalinity

Principle

The neutralizing capacity of a strong acid is defined as alkalinity of water. The presence of carbonates, bicarbonates along with hydroxyl ions makes water alkaline, various water may also consists phosphates, silicates and borates. These also contribute to impart alkalinity of water.

Apparatus

(*a*) Conical flask
(*b*) Burette
(*c*) Pipette
(*d*) Dropper

Reagents

(A) Hydrochloric Acid 0.1N (HCl)

Dissolve 8.33 ml. of concentrated HCl to 100 ml. distill water to prepare 1N HCl. Further dilute it 10 time to prepare 0.1N HCl (10ml $\rightarrow$ 100ml.)

(B) Methyl Orange Indicator

See CO_2.

(C) Phenolphthalein Indicator

See free CO_2.

Procedure

(*a*) Take 100 ml. of sample is a conical flask.
(*b*) Add 2-3 drops of phenolphthalein indicator.
(*c*) If solution remains colourless phenolphathalein alkalinity is absent.
(*d*) If solution turns pink coloured, titrate it with 0.1N HCl until solution becomes colourless (This is phenolphathlein alkalinity).
(*e*) Add 2-3 drops of methyl orange indicator.
(*f*) Yellow colour appears, continue titration with 0.1N HCl.
(*h*) At end point yellow colour changes to pink. (This is total alkalinity).

Calculation

$$\text{Phenolphthalein alkalinity (mg/l)} = \frac{(V1 - N \text{ of } HCL) \times 1000 \times 50}{\text{ml of sample taken}}$$

$$\text{Total alkalinity} = \frac{(V1 \times N \text{ of } NaOH) \times 1000 \times 50}{\text{ml of sample taken}}$$

where,

V: ml. of HCl with phenolphthalein

V_1: ml. of HCl with phenolphthalein and methyl orange.

Flow Diagram 4.6: Alkalinity Measurement

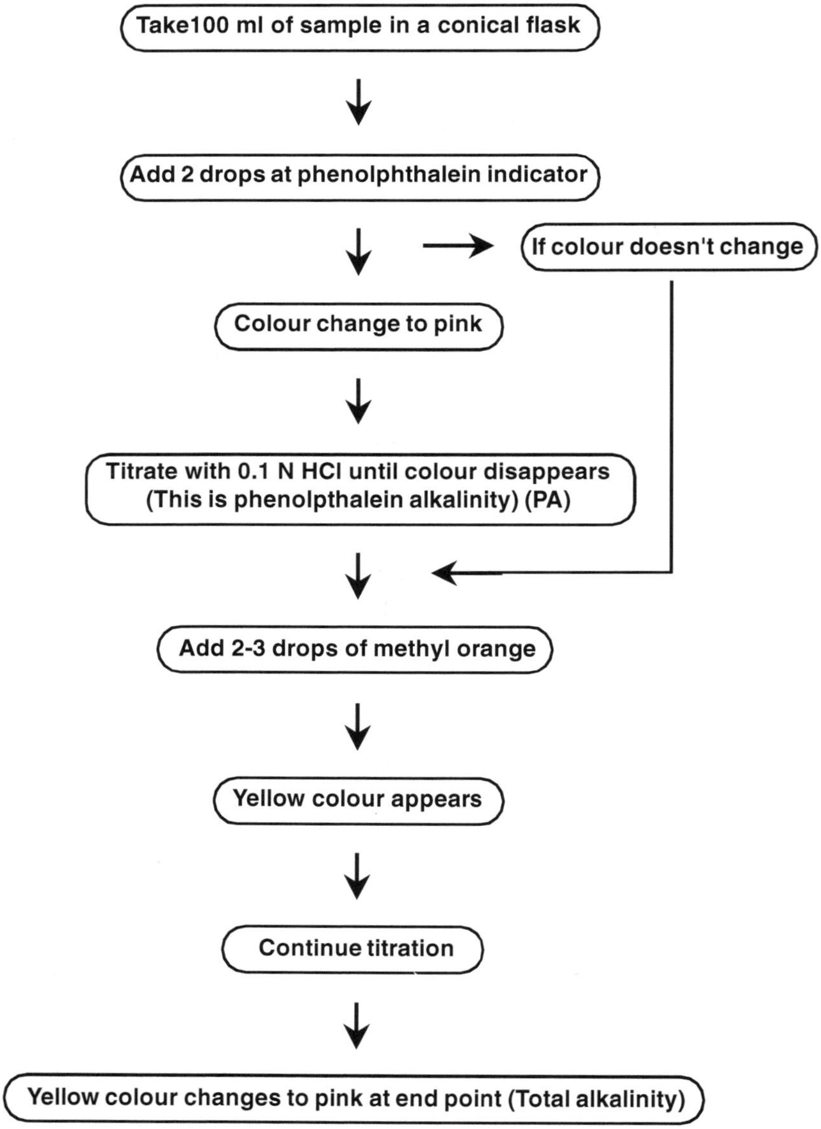

Sheet 4.6: Data Sheet for Estimation of Alkalinity

Sl.No.	Site/Source	Reading						Average	Standard Deviation
		A	B	C	D	E	F		
1.									
2.									
3.									
4.									
5.									
6.									
7.									
8.									
9.									
10.									

(h) Hardness

Principle

The capacity of water to form lather with soap is called the measure of water hardness. Mainly Ca and Mg ions are responsible for hardness of water however a minimal hardness of water is also produced by various polyvalent cations. Therefore total hardness may be defined as sum of Ca and Mg concentrations (as $CaCO_3$).

When EBT (Eriochrome Black T) an indicator is added to aqueous solution containing Ca & Mg it gives wine red colour. If this solution is titrated against EDTA. The complex of Ca & Mg are broken down and gives blue colour at end point.

Apparatus

(*a*) Conical flask
(*b*) Burette
(*c*) Pipette
(*d*) Dropper

Reagents

(A) Standard EDTA Solution (0.01 M)

Dissolve 3.723g of disodium salt of EDTA (Ethylene diamine tetra acetic acid) in distill water and finally make volume to one litre.

(B) Buffer Solution

(*a*) Dissolve 16.9 g of ammonium chloride (NH_4Cl) in 143 ml. of concentrated ammonium hydroxide (NH_4OH).

(*b*) Add 1.25 g of magnesium salt of EDTA.

In absence of magnesium salt of EDTA, dissolve 1.179g of disodium salt of EDTA and 780 mg $MgSO_4.7H_2O$ in 50ml. distilled water. Mix both (a) and (b) solutions and dilute to 250 ml. using distill water.

(C) Erichrome Black T

Mix 0.50 g of EBT indicator with 100g NaCl and grind it.

Procedure

(*a*) Take 50 ml. of sample in a conical flask. If sample have high calcium take 25 ml. sample and dilute it to 50 ml. using distilled water.

(*b*) Add 1ml. of buffer solution to give pH of sample to 10.0.

(*c*) Add sufficient amount (100-200mg) of EBT indicator, solution turns wine red.

(*d*) Titrate it against EDTA.

(*e*) At end point wine red colour changes to blue.

Calculation

$$Hardness\ mg/l = \frac{ml\ of\ EDTA\ used \times 1000}{ml\ of\ sample\ taken}$$

Flow Diagram 4.7: Hardness Measurement

Take 50 ml sample in a conical flask

↓

Add 1 ml of buffer solution

↓

Add 100-200 mg. EBT

↓

Solution turns wine red

↓

Titrate the content against EDTA solution

↓

At end point colour changes from wine red to blue

Sheet 4.7: Data Sheet for Estimation of Hardness

Sl.No.	Site/Source	Reading						Average	Standard Deviation
		A	B	C	D	E	F		
1.									
2.									
3.									
4.									
5.									
6.									
7.									
8.									
9.									
10.									

(i) Calcium

Principle

When EDTA is used to titrate water containing both calcium and magnesium. It first combines with calcium having great affinity towards it. As EDTA combines with both Ca and Mg an indicator is used that form a complex only with calcium such as ammonium purpurate. In this process magnesium in precipitated as hydroxide at higher pH.

Apparatus

(*a*) Conical flask
(*b*) Burette
(*c*) Pipette
(*d*) Measuring cylinder
(*e*) Dropper

Reagents

(A) Standard EDTA Solution 0.01M

See in Hardness.

(B) Sodium Hydroxide 1 N

Dissolve 20g of NaOH pellets in 500ml. distill water.

(C) Mureoxide (Ammonium Purpurate) Indicator

Mix 200mg of ammonium purpurate with 100gms. of NaCl and grind.

Procedure

(*a*) Take 50ml. of sample in conical flask. If the sample has high alkalinity take 20ml. sample and dilute to 50ml. using distilled water.

(*b*) Add 2.0 ml. of 1N NaOH solution.

(*c*) Now add sufficient quantity (100-200mg) of mureoxide indicator.

(*d*) A pink colour will appear.

(*e*) Titrate it against EDTA solution.

(*f*) At end point pink colour will change to purple.

Calculation

$$Calcium\ mg/l = \frac{Volume\ of\ EDTA\ used \times 400.8}{ml\ of\ sample\ taken}$$

Flow diagram 4.8 : Calcium measurement

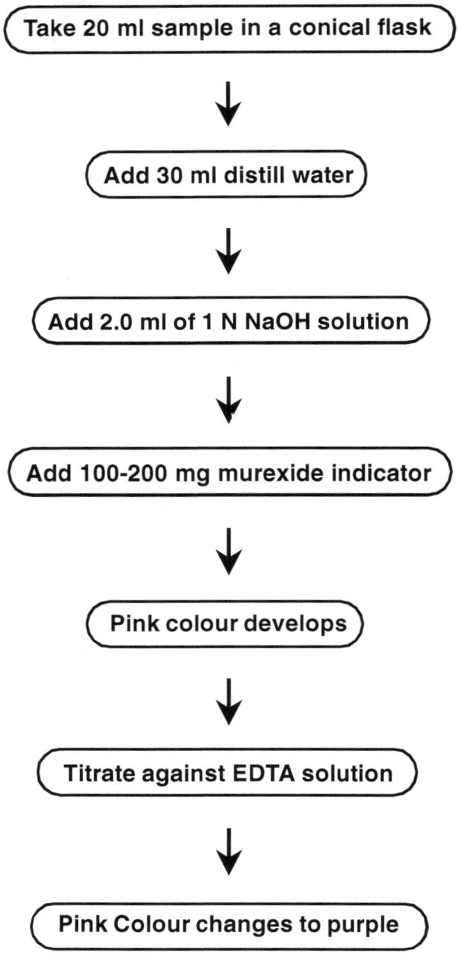

Sheet 4.8: Data Sheet for Estimation of Calcium

Sl.No.	Site/Source	Reading						Average	Standard Deviation
		A	B	C	D	E	F		
1.									
2.									
3.									
4.									
5.									
6.									
7.									
8.									
9.									
10.									

(j) Magnesium

Principle

EDTA has greater affinity to combine with calcium and magnesium but it combines first with calcium in presence of particular indicator. Therefore magnesium can be easily obtained by substracting calcium from both Ca & Mg concentration in water. In other words we can say that Mg is obtained by substracting Ca Concentration from total hardness.

Procedure

(*a*) Find out the volume of EDTA used for calcium concentration.

(*b*) Find out the volume of EDTA used for hardness determination.

Calculation

(a) $$Magnesium\ ions\ mg/l = \frac{(A-B)\times 400.8}{ml\ of\ sample\ taken}\times 1.645$$

where,

A: EDTA used in hardness determination.

B: EDTA used in Ca^{++} determination.

(b) Mg^{++} mg/l = Total hardness – Calcium hardness

Note: Always use equal quantity of sample in determining total hardness and calcium hardness for measuring magnesium.

Flow Diagram 4.9: Magnesium Measurement

Find volume of EDTA used in Calcium

↓

Find volume of EDTA use in Hardness

Sheet 4.9: Data Sheet for Estimation of Magnesium

Sl.No.	*Site/Source*	*Reading*						*Average*	*Standard Deviation*
		A	*B*	*C*	*D*	*E*	*F*		
1.									
2.									
3.									
4.									
5.									
6.									
7.									
8.									
9.									
10.									

(k) Chlorides

Principle

For chloride determination, silver nitrate is used as a titrant which has the affinity to react with chloride ions and forms Silver chloride (AgCl). The end point of titration is indicated by red colour produced by the reaction between silver nitrate and potassium chromate which is used as an indicator.

Apparatus

(*a*) Conical flask
(*b*) Burette
(*c*) Pipette
(*d*) Dropper
(*e*) Measuring cylinder

Reagents

(a) Silver Nitrate Solution 0.02 N

Dissolve 3.4 gm of silver nitrate in distill water and make the volume to one litre.

(b) Potassium Chromate Indicator

Dissolve 25gm of potassium chromate in 500ml. distill water.

Procedure

(*a*) Take 50ml. of sample in a conical flask.

(*b*) Add 1-2ml. of potassium chromate indicator.

(*c*) Titrate the sample with 0.02N Silver nitrate solution.

(*d*) At end point a perceptible red tinge colour appears.

Calculation

$$Chloride\ mg/l = \frac{(ml \times N\ of\ AgNO_3) \times 1000 \times 35.5}{ml\ of\ sample\ taken}$$

Flow diagram 4.10: Chlorides measurement

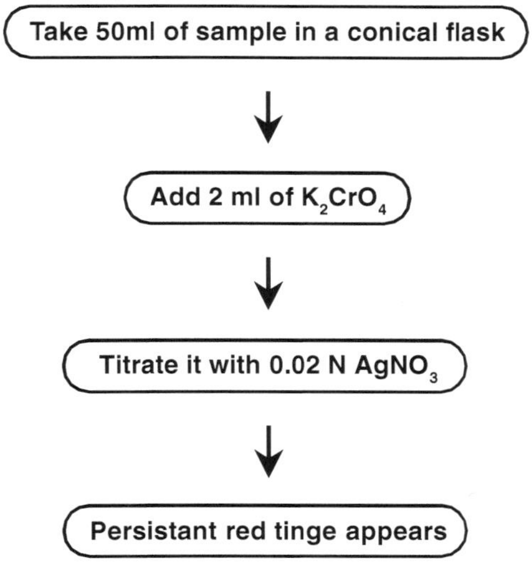

Sheet 4.10: Data Sheet for Estimation of Chloride

Sl.No.	Site/Source	Reading						Average	Standard Deviation
		A	B	C	D	E	F		
1.									
2.									
3.									

Contd...

Sheet 4.10–Contd...

Sl.No.	*Site/Source*	*Reading*						*Average*	*Standard Deviation*
		A	*B*	*C*	*D*	*E*	*F*		
4.									
5.									
6.									
7.									
8.									
9.									
10.									

(l) Phosphorus

(*i*) Inorganic Phosphorus

Molybdophosphoric acid is formed when ammonium molybdate is added to sample containing phosphorus, which in presence of stannous chloride gets reduced to molybednum blue giving appearance of blue colour. The absorption of light by this blue colour can be observed at 690 nm using spectrophotometer which gives the concentration of phosphate in sample.

Apparatus

(*a*) Conical Flask

(*b*) Measuring cylinder

(*c*) Pipette

(*d*) Dropper

(*e*) Spectrophotometer

Reagents

(a) Ammonium Molybdate Solution

(*i*) Dissolve 25 gms. of ammonium molybdate in 175 ml of distilled water. (*ii*) Add 280 ml of concentrated H_2SO_4 to 400 ml distilled water and allow it to cool. Mix (*i*) and (*ii*) and finally make its volume to one litre using distilled water.

(b) Stannous Chloride Solution

Dissolve 2.5 gms. of fresh stannous chloride in 100 ml glycerol. Heat and stir for rapid dissolution.

(c) Standard Phosphate Solution

Dissolve 219.5 mg of potassium hydrogen phosphate in distilled water and make up the volume to one litre.

$$1.0 \text{ ml} = 50.0\ \mu g PO_4^{3-}$$

Procedure

(*a*) Take 50 ml of sample in a conical flask.

(*n*) Add 2 ml of ammonium molybdate solution followed by 4-5 drops of stannous chloride solution.

(*c*) A blue colour will appear.

(*d*) Take reading at 690 nm on a Spectrophotometer.

(*e*) Run a blank with the same amount of chemicals.

(*f*) Compare the concentration using calibration curve.

Note: As blue colour appears take reading after 10 minutes but before 12 minutes use same time interval for all readings.

Flow Diagram 4.11: Phosphorus Measurement

Take 50 ml sample in a conical flask

↓

Add 2 ml ammonium molybdate solution followed by 4-5 drops of Stannous chloride solution

↓

Blue colour will appear

↓

Take reading at 690 nm using Spectrophotometer

Sheet 4.11: Data Sheet for Estimation of Inorganic Phosphorous

Sl.No.	*Site/Source*	*Reading*						*Average*	*Standard Deviation*
		A	*B*	*C*	*D*	*E*	*F*		
1.									
2.									
3.									
4.									
5.									
6.									
7.									
8.									
9.									
10.									

(*ii*) Total Phosphorous

Principle

Organic phosphorous is converted to orthophosphates by heating or by persulphate digestion while inorganic phosphates are converted to orthophosphates by H_2SO_4 digestion. The phosphates thus released can be determined colorometrically.

Apparatus

1. Conical flask
2. Beakers
3. Hot plate
4. Dropper

Reagents

A. **Sulphuric Acid**: Concentrated

B. **Potassium Sulphate**: Solid

C. **Sodium Hydroxide**: See Free CO_2

D. **Phenolphthalein Indicator**: See Free CO_2

E. **Ammonium Molybdate**: See inorganic phosphorus.

F. **Stannous Chloride Solution:** See inorganic phosphorus.

G. **Copper Sulphate:** Dissolve 10g $CuSO_4$ in 100ml. distill water.

Procedure

Digestion: H_2SO_4-K_2SO_4

Take 40ml sample in a conical flask add 4ml. concentrated H_2SO_4 and 10 drops of $CuSO_4$ now add 6g K_2SO_4 heat the flask over hot plate. During digestion the sample turns pale green, continue digestion for next half an hour. Replace flask, cool and finally make volume to 100ml.

Digestion: $H_2SO_4 - HNO_3$

Take 25 ml. of sample in a conical flask and add 1ml conc. H_2SO_4 and 5ml. of concentrated HNO_3. Digest it over hot plate till volume of sample becomes 1ml. replace flask when solution turns colourless.

Estimation

(*a*) Add 20ml. of distilled water to digest sample.

(*b*) Titrate the sample with 5N NaOH in case of $H_2SO_4 - K_2SO_4$ digestion while with 1N NaOH in case of $H_2SO_{4-}HNO_3$ digestion.

(*c*) At end point solution turns pink.

(*d*) Add 2ml. of ammonium molybdate solution followed by 4-5 drops of stannous chloride.

(*e*) A blue colour will appear.

(*f*) Take the reading at 690nmm using spectrophotometer.

(*g*) Compare concentration using standard curve.

Flow Diagram 4.12: Phosphorus Measurement

Take 25 ml of the sample in kjeldhal flask

↓

Add 1 ml of H_2SO_4 and 5 ml of Conc. HNO_3

↓

Digest the sample on a heater till vol. becomes nearly 1 ml

↓

Continue heating further until solution become colourless

↓

Add 20ml of distilled water

↓

Add 1 drop of phenolphthalein

↓

Titrate it with 1 N NaOH

↓

Pink colour will appear at end point

↓

Add 2 ml of ammonium molybdate

↓

Add 5 drops of $SnCl_2$

↓

A blue colour will appear

↓

Take reading at 690 nm

Sheet 4.12: Data Sheet for Estimation of Total Phosphorous

Sl.No.	Site/Source	Reading						Average	Standard Deviation
		A	B	C	D	E	F		
1.									
2.									
3.									
4.									
5.									
6.									
7.									
8.									
9.									
10.									

(m) Phenols

Principle

Phenols are absent in natural water and are found as a result of mixing of effluents from various industries. The concentrations of phenols in sample may vary depending upon the nature of sample.

Apparatus

1. Separating funnel
2. Beaker
3. Measuring cylinder
4. Spectrophotometer

Procedure

1. Take 50ml. sample in a beaker, adjusts its pH to 12 using NaOH.
2. Transfer the sample to a separating funnel.
3. Extract the sample using 20ml methylene chloride.
4. Allow it to stand for 10 min. Discard organic phase.
5. Again adjust the pH to 2 using 5N HCl.
6. Extract three times with methylene chloride using same amount (20ml.).
7. Discard aqueous phase and extract organic phase with 0.5N NaOH three times.
8. Take the aqueous phase and adjust the pH to 2 using 5N HCl.
9. Extract the aqueous phase using methylene chloride for three times.

10. Pooled organic phase and dry it over anhydrous sodium sulphate.
11. Completely evaporate it is ratovapour (30-35°C).
12. Residue was dissolved in methanol (2ml.)
13. Take reading at 460 nm using spectrophotometer and compare with standard curve.
14. Run a blank using same amount of chemical.

Calculation

$$\mu g\ Phenol/l = \frac{A}{B} \times 100$$

where,

A: mg phenol in sample from calibration curve

B: ml. of sample taken.

Flow Diagram 4.13: Phenol Measurement

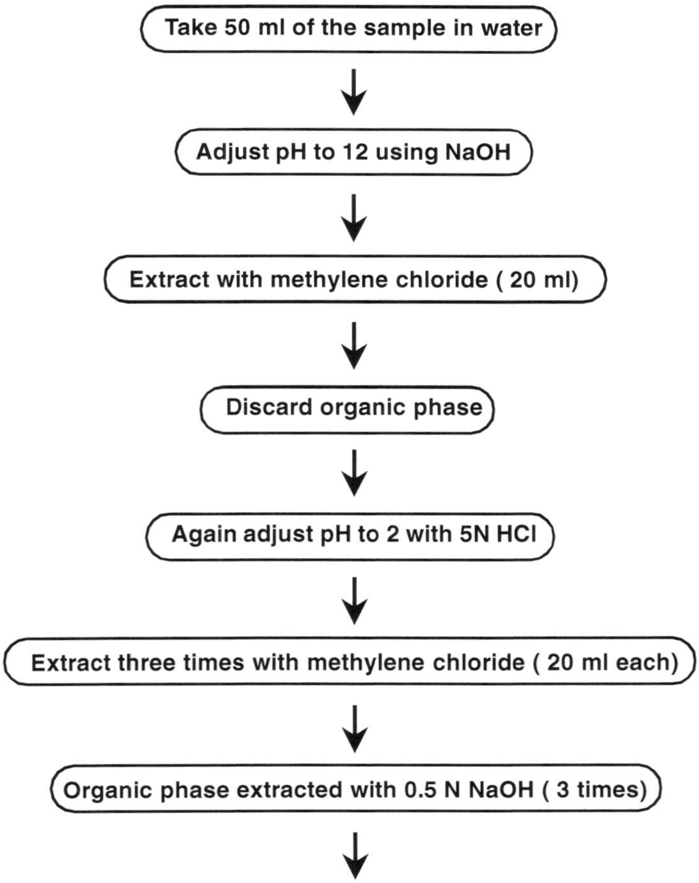

↓

Again adjust pH to 2 with 5N HCl

↓

Aqueous phase again extracted with methylene chloride (3 times)

↓

Pooled organic phase dried over anhydrous sodium sulphate

↓

Completely evaporate in rotavapour

↓

Residue was dissolved in methanol (2ml)

↓

Take reading at 460 nm using spectrophotometer

↓

Compare with standard curve

Sheet 4.13: Data Sheet for Estimation of Phenol

Sl.No.	*Site/Source*	*Reading*						*Average*	*Standard Deviation*
		A	*B*	*C*	*D*	*E*	*F*		
1.									
2.									
3.									
4.									
5.									
6.									
7.									
8.									
9.									
10.									

(n) Tannin & Lignin

Principle

Lignin is a plant constituent and is byproduct of pulp and paper industry. Therefore effluent from pulp and paper industries and tanning industries largely contributes tannin & lignin to water bodies.Both lignin and tannin are determined by their capacity to reduce folin phenol reagent (tungstophosphoric and molybdophosphoric acid) to form a blue colour with which we can estimate the concentration of tannin and lignin.

Apparatus

1. Spectrophotometer
2. Nessler tubes
3. Conical flask.

Reagents

A. Folinphenol Reagent

Available commercially prepared.

B. Carbonate Tartrate Reagent

Dissolve 20gm sodium carbonate (Na_2CO_3) and 1.2 g sodium tartrate ($(Na_2C_4H_4O_6).2H_2O$) in 75 ml hot distill water. Cool it and finally make its volume to 100ml.

C. Standard Solution

Dissolve 1g tannic acid, tannin or lignin to 1litre distill water. This is stock solution dilute 10ml stock solution to 1000ml. with distill water.

1.0ml = 10 mg

Procedure

1. Take 50ml. of clear sample.
2. Immediately add 1ml folin phenol reagent and 10ml carbonate-tartrate reagent.
3. Allow it to stand for about 30-40 minutes for colour development.
4. Compare either visually with blank or spectrophotemoterically against reagent blank prepared at the same time at 700 nm.

(B) Modified Pearl-Bensen (Nitrosation) method

This method is used to determine lignin content on effluent sample (Pearl Benson, 1940).

Reagents

(a) Acetic Acid (10%)

100ml. of acetic acid was taken and diluted to 1000 ml using distilled water.

(b) Sodium Nitrite (10%)

Dissolve 10g AR grade sodium nitrite in distilled water and make its volume to 100ml.

(c) Ammonium Hydroxide (1N)

Dissolve 135ml. of concentrated ammonium hydroxide (28%) as ammonia in distilled water and make volume to 1 litre.

Procedure

1. Take 50ml. of sample in conical flask.
2. Add 1ml. acetic acid and 1ml. sodium nitrite.
3. Mix the solution by stirring and allow it to stand for 20 min. at room temperature.
4. Now add 2ml. of ammonium hydroxide solution.
5. Mix the solution and allow it to stand for 10 min.
6. Read the absorbance at 430nm using spectrophotometer.
7. Run a blank of distilled water using same amount of chemicals.

Calculation

$$C = \frac{A_s}{ab}$$

where,

a: Absorptivity Cm^{-1} ppm^{-1} i.e = 0.000247Cm^{-1} ppm^{-1} (acc. to Device 1992)

As: Ne^{-1} absorbance

C: Concentration of lignin

b: Length of light path (1Cm).

Flow Diagram 4.14: Tannin & Lignin measurement

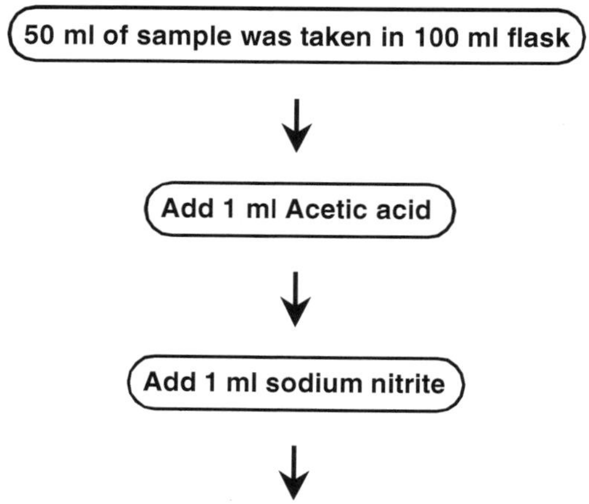

↓

Mix the solution by swirling & allowed to stand for 15 min. at room temperature

↓

Add 2 ml ammonium hydroxide, mix & allow the sol. to stand for 10 min. at room temp.

↓

Read absorbance at 430 nm using spectrophotometer

Sheet 4.14: Data Sheet for estimation of Tanin and Lignin

Sl.No.	Site/Source	Reading						Average	Standard Deviation
		A	B	C	D	E	F		
1.									
2.									
3.									
4.									
5.									
6.									
7.									
8.									
9.									
10.									

(o) Oil and Grease

Principle

Oil and grease are dissolved in suitable solvent and can be separated by using separating funnel. The solvent layer is evaporated and the residue will yield the oil and grease.

Apparatus

1. Separating funnel
2. Water bath
3. Filter paper

Reagents

A. Sulphuric Acid (1+1)

Acid 250ml. concentrated H_2SO_4 to 250 ml. distilled water.

B. Petroleum Ether

Boiling range 40°C-60°C.

Procedure

1. Take 250ml. of sample in a suitable separating funnel.
2. Add 5ml. of sulphuric acid and 20ml. of petroleum ether.
3. Wait for 5min. to separate the two layers.
4. Discard lower layer of sample.
5. Transfer the upper layer into a preweighted beaker through paper (filter paper must be wetted with petroleum ether).
6. Continue extraction 2-3 times and add it into pre weighted beaker.
7. Evaporate the petroleum ether on a water bath and take final weight of beaker after cooling.

Calculation

$$\text{Oil and Grease mg/l} = \frac{(W_1 - W) \times 1000}{V}$$

where,

W_1: Final weight of dish
W: Initial weight of dish
V: Volume of sample taken.

Flow Diagram 4.15 : Oil & Grease Measurement

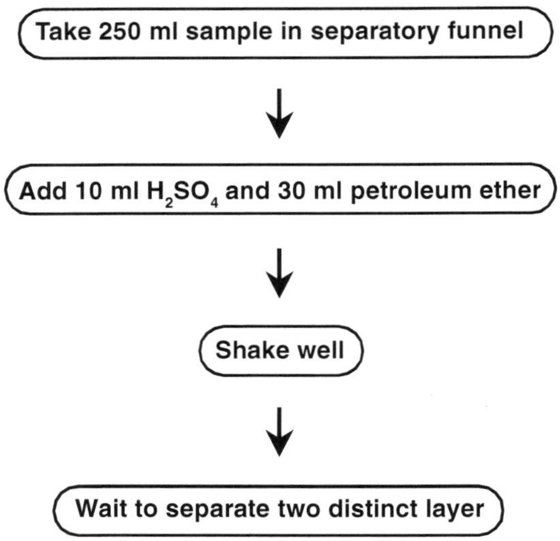

↓

Discard lower layer

↓

Filter petroleum ether into preweighted dish from separatory funnel

↓

Evaporate petroleum ether through water bath

↓

Weight dish

Sheet 4.15: Data Sheet for Estimation of Oil and Grease

Sl.No.	*Site/Source*	*Reading*						*Average*	*Standard Deviation*
		A	*B*	*C*	*D*	*E*	*F*		
1.									
2.									
3.									
4.									
5.									
6.									
7.									
8.									
9.									
10.									

(p) Sulphate

(1) Gravimetric Method

Principle

In presence of HCl, sulphate is precipitated as Barium sulphate ($BaSO_4$) by adding Barium chloride ($BaCl_2$). The precipitate is washed with water to remove Cl^- filtered dried and weighed as $BaSO_4$.

Apparatus

1. Steam bath
2. Oven
3. Analytical balance
4. Filter paper.

Reagent

A. Methyl Orange Indicator

Dissolve 100mg methyl red in 100ml. distilled water.

B. Hydrochloric Acid

1+1.

C. Barium Chloride Solution

Dissolve 100g $BaCl_2.2H_2O$ in 1 liter distilled water, filter the solution before use.

D. Silver Nitrate-nitric Acid Reagent

Dissolve 8.5g of $AgNO_3$ and 0.5 ml. concentrated HNO_3 in 500ml. distilled water.

Procedure

1. If sample contains suspended impurities, evaporate suitable volume of sample to dryness in a platinum dish on steam bath.
2. Add 1ml. HCl, continue evaporation to dryness.
3. Add 2ml. distilled water and 1ml. HCl (1+ 1) and evaporate to dryness over steam bath.
4. Add 2ml. HCl take residue is hot water and filter.
5. Wash filter paper several times with hot distilled water.
6. Combine filterate and washing, discard residue.
7. Make the volume to 200ml. adjust pH to 4.5-5.0 with HCl or using methyl red.
8. Now add 2ml. HCl, boil solution and add $BaCl_2$, solution slowly until precipitation appears to be completed.
9. Digest solution at 80-90°C for more than 2 hrs.
10. Filter the precipitate and wash it with distilled water until it becomes free of Cl^- by testing it with $AgNO_3$–HNO_3 reagent.
11. Dry precipitate and ignite it at 800°C for 1 hr.
12. Cool and weigh it.

Calculation

$$Sulphate\ (SO_4)\ mg/l = \frac{mg\ BaSO_4 \times 411.50}{ml\ of\ sample\ taken}$$

Flow Diagram 4.16: Sulphate measurement (Method - 1)

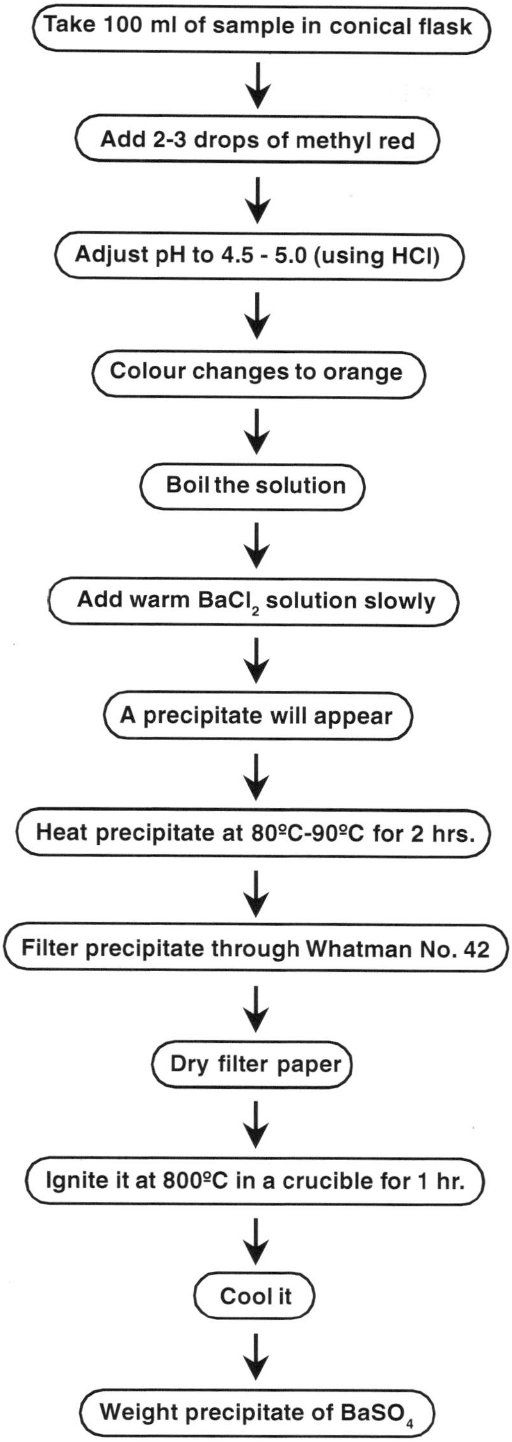

Sheet 4.16: Data Sheet for Estimation of Sulphate

Sl.No.	*Site/Source*	*Reading*						*Average*	*Standard Deviation*
		A	*B*	*C*	*D*	*E*	*F*		
1.									
2.									
3.									
4.									
5.									
6.									
7.									
8.									
9.									
10.									

(2) Turbidimetric Method

Principle

Sulphate is precipitated as Barium sulphate in acidic medium using barium chloride.The concentration of sulphate should be measured by absorbance of light by the sample and compare it with standard curve.

Apparatus

1. Magnetic stirrer
2. Spectrophotometer
3. Conical flask
4. Measuring cylinder

Reagents

A. Conditioning Reagent

Dissolve 75g of NaCl, 30ml. of concentrated. HCl, 100ml. 95% ethyl alcohol in about 300 ml. distilled water. Now add 50ml. glycerol and mix it.

B. Barium Chloride

Solid.

C. Standard Sulphate Solution

Dilute 0.147mg anhydrous Na_2SO_4 in one litre distilled water. 1.0 ml = 100 $\mu g SO_4^{2-}$.

Procedure

1. Take 100ml. of sample is conical flask.
2. Add 5ml. conditioning reagent and stir the sample oven magnetic stirrer.
3. While stirring add a spoonful of $BaCl_2$, stir for one more minute.
4. Take the reading on spectrophotometer at 420 nm.
5. Prepare the standard curve, for 0.0-40.0 mg/lt. at the interval of 5mg/lt. above 40mg/lt. its accuracy decreases.

Calculation

$$Sulphate\ mg/l = \frac{BaSO_4 \times 1000}{ml\ of\ sample\ taken}$$

Flow Diagram 4.17: Sulphate Measurement (Method-2)

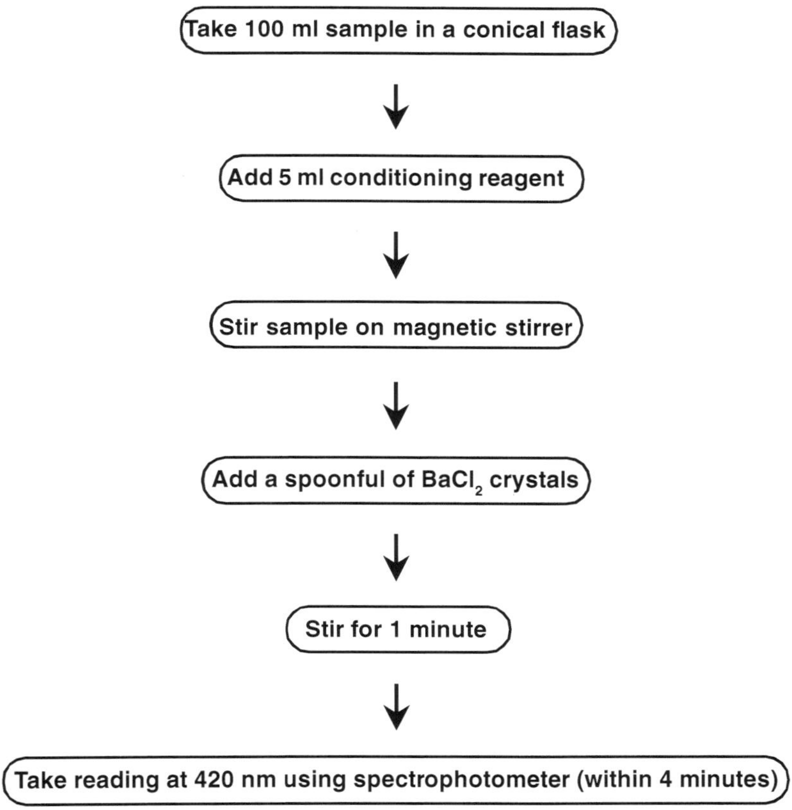

Sheet 4.17: Data Sheet for Estimation of Sulphate

Sl.No.	*Site/Source*	*Reading*						*Average*	*Standard Deviation*
		A	*B*	*C*	*D*	*E*	*F*		
1.									
2.									
3.									
4.									
5.									
6.									
7.									
8.									
9.									
10.									

(q) Sodium

(1) Flame Photometric Method

Principle

Sodium can be determined by using flame photometer at 589nm. When sample is produced in flame it comes to excitation condition and produces characteristic light. The light intesity at 589nm is proportional to concentration of sodium which can be read by using light dispersion devices.

Apparatus

1. Flame Photometer
2. Conical flask
3. Measuring cylinder

Reagents

A. Stock Sodium Solution

Dissolve 2.542 g of NaCl (previously dried at 140°C) in distilled water and make volume to one litre (1ml = 1.00 mg Na)

B. Intermediate Sodium Solution

Dissolve 10ml. of stock solution in distilled water and make volume to 100ml. (1 ml. = 0.10 mgNa)

C. Nitric Acid

Concentrated.

D. Hydrochloric Acid

Concentrated.

E. Hydrogen Peroxide

30%.

F. Ammonium Hydroxide

Concentrated.

Procedure

1. For waste water and highly polluted samples pretreat samples as follows:

   Take suitable quantity of sample in a conical flask, add nitric acid and evaporate it on waterbath further add 25ml. HNO_3 and evaporate until sample volume becomes nearly 2ml. Presence of brown fumes indicates unoxidized organic matter add more HNO_3 and H_2O_2 for complete ashing of organic matter. When sample becomes colourless, dissolve in small amount of HCl and distilled water filter it, neutralize with concentration NH_4OH and dilute for further process.
2. Follow instructions of manufacturer manual for operating instrument and take readings.
3. Prepare calibration curve in the range of 0-1, 0 to 10 or 0-100 mg/l. using sodium standard solution.

Calculation

Sodium mg/l = (mg Na/l in sample) × dilution ratio

Flow Diagram 4.18: Sodium Measurement

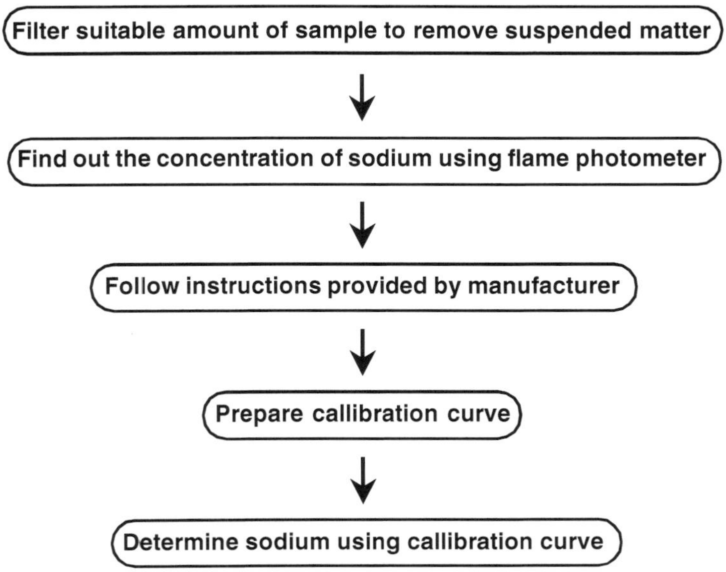

Sheet 4.18: Data Sheet for Estimation of Sodium

Sl.No.	Site/Source	Reading						Average	Standard Deviation
		A	B	C	D	E	F		
1.									
2.									
3.									
4.									
5.									
6.									
7.									
8.									
9.									
10.									

(r) Potassium

(1) Flame Photo Metric Method

Principle

Potassium concentration can directly be determined using flame photometer at 766.5 nm. The intensity can be read on a scale using filter.

Apparatus

As is sodium.

Reagents

A. Stock Potassium Solution

Dissolve 1.907g of KCl (previously dried at 110°C) in distilled water and make volume to one liter. (1ml=1.0 mg K)

B. Intermediate Potassium Solution

Dissolve 10ml stock K solution and make volume to 100ml using distilled water (1ml = 0.100 mg K)

C. Standard Potassium Solution

Dissolve 10ml intermediate K solution and make it 100ml with distilled water (1ml = 0.01 mg K).

Procedure

Same as in Na, but measure emission intensity at 766.5 nm.

Calculation

$$K\,mg/lt = (mg\,K/l\ is\ sample) \times dilution\ ratio$$

Sheet 4.19: Data Sheet for Estimation of Potassium

Sl.No.	Site/Source	Reading						Average	Standard Deviation
		A	B	C	D	E	F		
1.									
2.									
3.									
4.									
5.									
6.									
7.									
8.									
9.									
10.									

(s) Nitrate

(1) Brucine Method

Principle

Nitrate when reacts with brucine produces yellow colour is presence of H_2SO_4 this can be measured spectrophotometerically at 410nm.

Apparatus

1. Spectrophotometer
2. Test tubes.

Reagents

A. Nitrate Stock Solution

Dissolve 722mg of KNO_3 in distilled water and make volume to 1 liter (1 ml. = 0.1 mg Nitrate)

B. Standard Nitrate Solution

Dilute 100 ml. of stock solution in distilled water and make volume to 1 liter (1 ml. = 0.01 mg Nitrate)

C. Brucine - Sulphanilic Acid Solution

Dissolve 1gm brucine sulphate and 0.1g sulphanilic acid in 70ml. hot distilled water. Add 3ml. concentrated HCl cool the solution and dilute it to 100ml. with distilled water.

D. Sulphuric Acid Solution

Add 500ml. concentrated H_2SO_4 in 125 ml. distilled water and cool it.

E. Sodium Chloride Solution

Dissolve 300gm NaCl in distilled water and make volume to 1 liter.

F. Sodium Arsenite Solution

Dissolve 1.83gm Sodium arsenite ($NaAsO_2$) in 100ml. distilled water.

Procedure

1. If sample consists of residual chlorine remove it by adding 0.05ml of sodium arsenite solution. Add one more drop and mix well.
2. Take 10 ml. of sample in a test tube and add 2ml. of NaCl solution.
3. Now add 10ml. H_2SO_4 solution after mixing contents.
4. Add 0.5 ml. brucine Sulphanilic acid and mix thoroughly.
5. Place tube in water bath for 15-20 minutes.
6. Cool the tube and measure the reading at 410 nm using spectrophotometer.
7. Find out the concentration using standard curve.
8. Prepare standard curve between concentration and absorbance by taking dilutions from 0.1-1.0 mg/l using same amount of chemicals as taken for sample.

Flow Diagram 4.19: Nitrate Measurement

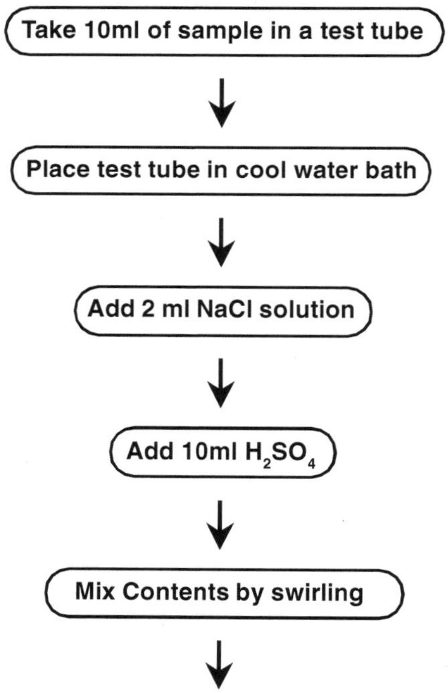

↓

Add 0.5 ml brucine reagent

↓

Place the tube in hot water bath for 30 min.

↓

Cool the tube

↓

Take reading at 410 nm using spectrophotometer

Sheet 4.20: Data Sheet for Estimation of Nitrate

Sl.No.	*Site/Source*	*Reading*						*Average*	*Standard Deviation*
		A	*B*	*C*	*D*	*E*	*F*		
1.									
2.									
3.									
4.									
5.									
6.									
7.									
8.									
9.									
10.									

(t) Nitrite

Principle

In an acidic medium of 2-2.5 pH nitrite form a diazonium salt with sulphanilic acid which is coupled with α-naphthylamine hydrochloride and forms a pinkish coloured dye.

Apparatus

1. Spectrophotometer
2. Conical flask.

Reagents

A. EDTA Solution

Dissolve 500mg. of disodium salt of EDTA in distilled water and make volume to 100ml.

B. Sulphanilic Acid Solution

Dissolve 600mg sulphanilic acid in 70ml. of hot distilled water and add 20 ml. concentrated HCl. Finally dilute it to 100ml using distilled water.

C. α-naphthylamine Hydrochloride Solution

Dissolve 600 mg of α-naphthylamine hydrochloride solution and 1ml. concentrated HCl in distilled water and make volume to 100ml. Place it is refrigerator.

D. Sodium Acetate Solution

Dissolve 16g of anhydrous CH_3COONa in distilled water and make volume to 100 ml.

E. Standard Nitrite Solution

Dissolve 1.232g of $NaNO_2$ in distilled water and make volume to one litre. Dilute this stock solution to 250 times *i.e.* 2ml $\rightarrow$ 500ml. This is standard solution containing 1mg/l Nitrite.

Procedure

1. Take 50ml. of sample is a conical flask.
2. Add 1ml. EDTA solution, 1ml. Sulphuric acid and 1ml. of α-naphthylamine hydrochloride solution and 1ml. of sodium acetate in sequence.
3. A wine red colour will appear
4. Take reading at 520 nm using spectrophotometer.
5. Compare the concentration with standard curve to find nitrite.
6. Prepare the standard curve between 0-1mg nitrite/l at interval of 0.1 using same amount of chemicals as taken for sample.

Flow Diagram 4.20: Nitrite Measurement

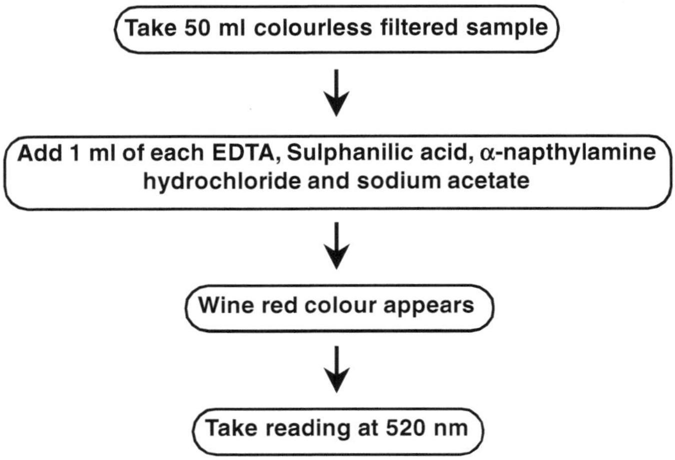

Sheet 4.21: Data Sheet for Estimstion of Nitrite

Sl.No.	Site/Source	Reading						Average	Standard Deviation
		A	B	C	D	E	F		
1.									
2.									
3.									
4.									
5.									
6.									
7.									
8.									
9.									
10.									

(u) Flourides

Principle

Flourides plays an important role in water, they are found in varying amount and are responsible for dental problem such as motled enamel or Flourosis. The quantity of flouride in water should be in concentration ranging between 0.5 to 1.0.

Calculation

$$\text{Flouride mg/l} = \frac{A}{\text{ml of sample taken}} \times \frac{B}{C}$$

where,

A: µg Flouride determined photometrically

B/C: Only when sample is distilled, B/C applies to a volume B, and a portion C taken from it for colour development.

Flow Diagram 4.21: Flouride Measurement

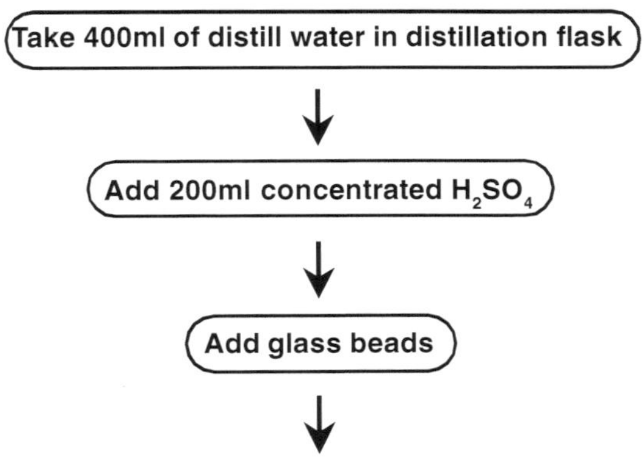

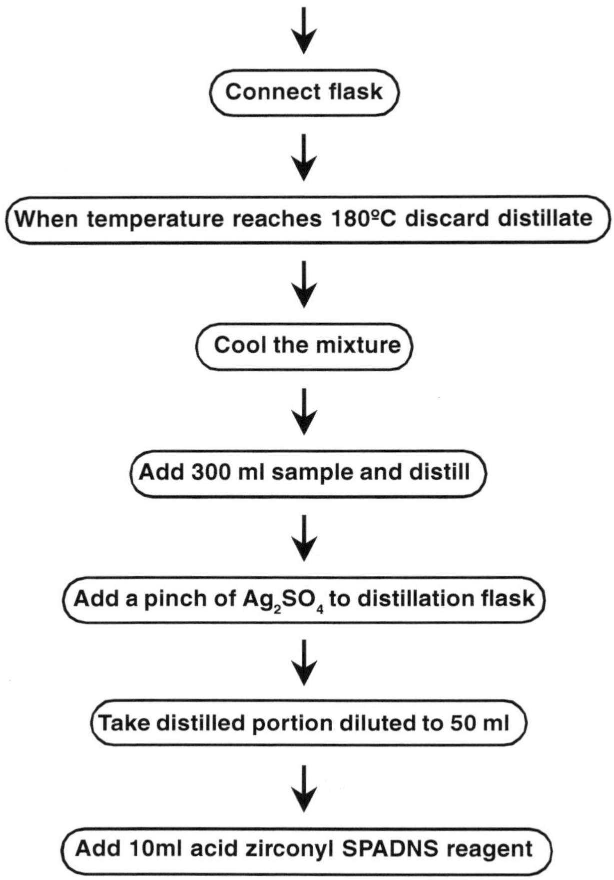

Sheet: 4.22: Data Sheet for Estimation of Fluoride

Sl.No.	*Site/Source*	*Reading*						*Average*	*Standard Deviation*
		A	*B*	*C*	*D*	*E*	*F*		
1.									
2.									
3.									
4.									
5.									
6.									
7.									
8.									
9.									
10.									

(v) Oxidation Reduction Potential (Eh)

Principle

A substance which losses electrons is oxidized and the substance, which gains, is reduced. When a platinum electrode is immersed in such a system, a potential is developed on it depending upon the ratio of oxidized and reduced states, is called as redox potential. A solution containing an excess of the reduced state will be oxidized after losing its electrons to platinum; while if solution contains more oxidized state will be reduced, after taking the electrons from the platinum electrode. This potential can be measured by using a reference electrode. Eh is determined indirectly from the equation based on pH of water sample, *i.e.*

$$Eh = Eo - 0.058\ pH + 0.0145 \times \log pO_2$$

where,

Eo: Standard electrode potential is determined directly from Figure 4.1.

pO_2: Partial oxygen pressure of water system = dissolved oxygen.

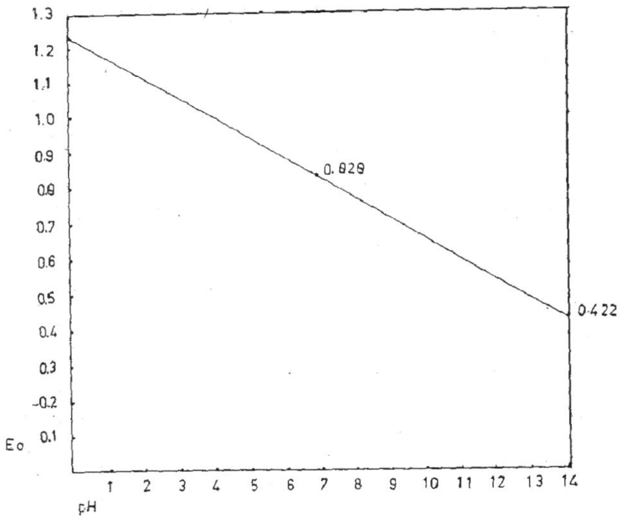

Figure 4.1: Standard Electrode Potential

(x) Oxidation Reduction Index (rH_2)

It is calculated by the equation:

$$rH_2 = \frac{Eh}{0.029} + 2pH$$

Chapter 5
Heavy Metals and its Analysis

(A) Chromium

Principle

Oxidize the sample with potassium permagnate so that all the chromium must be converted to hexavalent chromium because this colorimetric method measures only hexavalent chromium. Now determine this hexavalent chromium colometrically by reacting it with diphenyl-carbazide in an acidic condition, a reddish violet colour is produced.

Apparatus

1. Spectrophotometer
2. Beakers
3. Hot plate
4. pH meter.

Reagent

A. Stock Chromium Solution

Dissolve 141.4 mg $K_2Cr_2O_7$ in distilled water and dilute it to 1 litre (1.00 ml–50mg Cr).

B. Standard Chromium Solution

Dilute 100 ml. stock solution in distilled water and make volume to 100ml. (1.00ml–5 mg Cr.)

C. Nitric Acid

Concentrated.

D. Sulphuric Acid

Concentrated.

E. Methyl Orange Indicator

See Free CO_2

F. Hydrogen Peroxide

H_2O_2 (30%)

G. Ammonium Hydroxide

NH_4OH Concentrated.

H. Potassium Permagnate Solution

Dissolve 4gm $KMnO_4$ in 100ml. distilled water.

I. Sodium Azide Solution

Dissolve 0.5g sodium azide in 100ml. distilled water.

J. Diphenyl Carbazide Solution

Dissolve 250mg of 1.5 diphenyl carbazide in 50ml. acetone and store it in brown bottle. Discard after discolouration.

K. Chloroform

$CHCl_3$

L. Phosphoric Acid

H_3PO_4 Concentrated.

M. Sulphuric Acid 0.2 N

Dilute 17 ml. of 6NH_2SO_4 to 500ml. distilled water.

N. Cupferron Solution

Dissolve 5gm of cupferron in 95ml. distilled water.

Procedure

Nitric Acid-Sulphuric Acid Digestion

1. Take 50ml. of sample in a beaker, if solution is not already acidified, acidify to methyl orange end point using H_2SO_4.
2. Now add 5ml. concentrated HNO_3 and few glass beads.
3. Bring to slow boil on hot plate and evaporate to 15-20 ml.
4. Now add 5ml. concentrated HNO_3 and 10ml. of concentrated H_2SO_4 evaporate until dense fumes of SO_3 first appears.
5. If solution is not clear, add 10ml. more HNO_3 & repeat evaporation.

6. Heat to remove all HNO_3, when solution becomes clear and no brownish fumes are evident. (Never let the sample dry during digestion).
7. Cool and dilute it to make 50ml. with distilled water.
8. Heat it to boiling point so that all soluble salts may dissolve.
9. Filter the solution if necessary & transfer it in a volumetric flask.
10. Cool and dilute it to 100ml. & mix thoroughly.

Total Dissolved Chromium

1. Take suitable amount of sample (containing 10-100 µg/ltr.) in a 125 ml. separating funnel.
2. Dilute it to 40ml. with distilled water and cool it over ice bath.
3. Add 5ml. of cupferron solution, shake and keep it over ice bath for 1min.
4. Extract the sample in separating funnel with three successive portrain of $CHCl_3$ (5ml. each).
5. Let layers separate after shaking, discard $CHCl_3$ extract.
6. Transfer the aqueous phase in a conical flask.
7. Wash separating funnel with little amount of water and add this water to flask.
8. Boil for 5-8 minutes to evaporate $CHCl_3$.
9. Add 5ml. HNO_3 and sufficient H_2SO_4.
10. Boil over hot plate when SO_3 fumes appears. Add 5ml. more HNO_3.
11. Boil for complete decomposition of organic matter and to eliminate HNO_3.
12. Cool and add 25ml. distilled water.

Oxidation & Trivalent Chromium

1. Take suitable amount of sample (containing 10-100 mg/Cr) in a conical flask.
2. Add few drops of methyl orange as an indicator. Now slowly add NH_4OH until solution becomes basic to methyl orange.
3. Add (1N) H_2SO_4 dropwise until solution becomes acidic. Add 1ml. more in excess.
4. Make volume to 40ml. and boil it.
5. Now add 2 drops of $KMnO_4$ or more dropwise until dark red colour appears.
6. Boil for 2 minutes and add 1ml. NaN_3, continue boiling.
7. Add 1ml. NaN_3 if red colour doesn't fade after 35 sec.
8. Continue boiling until the colour fades.

Colour Development & Measurement

1. Take the sample & adjust pH to 1.0 ± 0.3 using 0.2 N H_2SO_4.
2. Dilute the solution to 100ml. and add 2ml. diphenyl carbazide solution.

3. Mix it well and allow it to stand for 10 minutes for colour development.
4. Measure the absorbance of sample at 540nm using spectrophotometer.
5. Use distilled water as reference.
6. Compare the concentration with calibration curve.

Preparation of Calibration Curve

Take measured amount of standard chromium solution ranging from 2-20ml. in a conical flask. Follow the steps as taken for determination of sample. Prepare a standard curve and compare with that of sample.

Calculation

$$Chromium\ mg/l = \frac{\mu g\ Cr\ from\ standard\ curve}{ml\ of\ sample\ digested \times ml\ portion\ from\ 100\ ml\ total\ digest} \times 100$$

Flow Diagram 5.1: Chromium Measurement (Method-1)
Nitric Acid-Sulphuric Acid Digestion

Take 50ml. of sample in a beaker, if solution is not already acidified, acidify to methyl orange end point using H_2SO_4

↓

Now add 5ml. concentration HNO_3 and few glass beads

↓

Bring to slow boil on hot plate and evaporate to 15-20 ml.

↓

Now add 5ml. concentrated HNO_3 and 10ml. of concentrated H_2SO_4 evaporate until dense fumes of SO_3 first appears

↓

If solution is not clear, add 10ml. more HNO_3 & repeat evaporation

↓

Heat to remove all HNO_3, when solution becomes clear and no brownish fumes are evident. (Never let the sample dry during digestion)

↓

↓

Cool and dilute it to make 50ml. with distilled H_2O

↓

Heat it to boiling point so that all soluble salts may dissolve

↓

Filter the solution if necessary & transfer it to volumetric flask

↓

Cool & dilute it to 100ml. & mix thoroughly

Flow Diagram 5.2: Total Dissolved Chromium

Take suitable amount of sample (containing 10-100 mg/l) in a 125 ml. separating funnel

↓

Dilute it to 40ml. with distilled H_2O and cool it over ice bath

↓

Add 5ml. of cupferron solution, shake & keep it over ice bath for 1min.

↓

Extract the sample in separating funnel with three successive portion of $CHCl_3$ (5ml. each)

↓

Let layers separate after shaking, discard $CHCl_3$ extract

↓

↓

Transfer the aqueous phase in a conical flask

↓

Wash separating funnel with little amount of water and add this water to flask

↓

Boil for 5-8 minutes to evaporate $CHCl_3$

↓

Add 5ml. HNO_3 and sufficient H_2SO_4

↓

Boil over hot plate when SO_3 fumes appears. Add 5ml. more HNO_3

↓

Boil for complete decomposition of organic matter to eliminate HNO_3

↓

Cool & add 25ml. distilled H_2O

Flow Diagram 6.3: Oxidation & Trivalent Chromium

Take suitable amount of sample (containing 10-100 mg/Cr) in a conical flask

↓

Add few drops of methyl orange as an indicator, Now slowly add NH_4OH until solution becomes basis to methyl orange

↓

↓

Add (1+1) H_2SO_4 drop-wise until solution becomes acidic. Add 1ml. more in excess

↓

Make volume to 40ml. & boil it

↓

Now add 2 drops of $KMnO_4$ or more dropwise until dark red colour appears

↓

Boil for 2 min. & add 1ml. NaN_3, continue boiling

↓

Add 1ml. NaN_3 if red colour doesn't fade after 35 sec

↓

Continue boiling until the colour fades then add 0.25 ml. H_3PO_4

Flow Diagram 5.4: Colour Development & Measurement

Take the sample & adjust pH to 1.0±0.3 using 0.2 N H_2SO_4

↓

Dilute the solution to 100ml. & add 2ml. diphenyl carbazide solution

↓

Mix it well & allow it to stand for 10 min. for colour development

↓

↓

Measure the absorbance of sample at 540nm using spectrophotometer

↓

Use distilled H_2O as reference

↓

Compare the concentration with calibration curve

Sheet 5.1: Data sheet for Estimation of Chromium

Sl.No.	Site/Source	Reading						Average	Standard Deviation
		A	B	C	D	E	F		
1.									
2.									
3.									
4.									
5.									
6.									
7.									
8.									
9.									
10.									

(B) Mercury

Principle

When dithizone is added to the sample containing mercury, an orange red colour forms at a pH of 1.0. The absorbance of sample under spectrophotometer gives concentration of mercury. However this method is not best for detection of mercury as it is reported that this method does not give accurate results. Hence best method for detection of mercury is by using AAS.

Apparatus

1. Spectrophotometer
2. Conical flask
3. Separating funnel

Reagents

A. Stock Mercury Solution

Dilute 135.4 mg $HgCl_2$ in 700 ml distilled water. Add 1.5ml concentrated HNO_3 & make the volume to 1 liter using distilled water.

B. Standard Mercury Solution

Dilute 10 ml stock solution to 1 liter with distilled water (1 ml =1 mg Hg).

C. Sulphuric Acid

Concentrated.

D. Potassium Permagnate Solution

Dissolve 5 gm $KMnO_4$ in 100ml distilled water.

E. Potassium Persulphate Solution

Dissolve 5 gm $K_2S_2O_8$ in 100ml distilled water.

F. Hydroxylamine Hydrochloride Solution

Dissolve 50 gms $NH_2OH.HCl$ in 100 ml distilled water.

G. Dithizone Solution

Dissolve 6mg dithizone in 100ml chloroform.

H. Sulphuric Acid 0.25 N

Dilute 6.9 ml concentrated H_2SO_4 to 1 liter distilled water.

I. Potassium Bromide Solution

Dissolve 40 gm KBr in 100 ml distilled water.

J. Phosphate Carbonate Buffer

Dissolve 150g Na_2HPO_4. 12 H_2O & 38g potassium carbamate in 1 liter distilled water. Purify it by extractng it with 10ml dithizone until last extract remains green.

Procedure

1. Take 500ml of sample in a conical flask.
2. Add 1ml $KMNO_4$ and 10ml concetrated H_2SO_4, stir & boil.
3. Add 5ml potassium per sulphate after 5-10 minutes and allow it to cool for 40 minutes.
4. If pink colour develops, add 1 drop or more of hydroxylamine - hydrochloride solution to remove it.
5. Extract this sample with 25 ml of dithizone in a separating funnel until last extract remains green.
6. Collect $CHCl_3$ extract in different funnel.

7. Add 50ml of 0.25N H_2SO_4 & 10ml KBr solution. in each separating funnel.
8. Wait for separating of layers.
9. Discard organic layer which consists of other dithizonates.
10. Wash aqueous phase with chloroform & discard $CHCl_3$ layer.
11. Now add 12ml phosphate -carbonate buffer and 10ml dithizonate solution.
12. Collect the organic layers containing mercury dithizonate & measure absorbance at 490nm.
13. Prepare a calibration curve using standard mercury solution following the similar steps.

Concentration of standard solution should be taken from 0.0ml -10ml diluted to 500 ml.

Flow Diagram 5.4: Mercury Measurement

Take 500ml of sample in a conical flask

↓

Add 1ml $KMNO_4$ & 10ml conc. H_2SO_4, stir & boil

↓

Add 5ml potassium per sulphate after 5-10 min. & allow it to cool for 40 min.

↓

If pink colour develops, add 1 drop or more of hydroxylamine-hydrochloride solution to remove it

↓

Extract this sample with 25 ml of dithizone in a separating funnel until last extract remains green

↓

Collect $CHCl_3$ extract in different funnel

↓

Add 50ml of 0.25N H_2SO_4 & 10ml KBr solution in each separating funnel

↓

Wait for separating of layers

↓

Discard organic layer which consists of other dithizonates

↓

Wash aqueous phase with chloroform & discard $CHCl_3$ layer

↓

Now add 12ml phosphate-carbonate buffer & 10ml dithiozone solution

↓

Collect the organic layers containing mercury dithizonate & measure absorbance at 490mm

Sheet 5.2: Data Sheet for Estimation of Mercury

Sl.No.	*Site/Source*	*Reading*						*Average*	*Standard Deviation*
		A	*B*	*C*	*D*	*E*	*F*		
1.									
2.									
3.									
4.									
5.									
6.									
7.									
8.									
9.									
10.									

(C) Arsenic

Principle

In an acidic medium arsenic is reduced to arsine in presence of zinc which is further observed in to silver diethyl dithio carbamate solution which forms red colour and is measured under spectrophotometer at 535 nm.

Apparatus

1. Arsenic esimation apparatus.
2. Spectrophotometer

Reagents

A. HCl

Concentrated

B. Zinc

20-30 mesh.

C. Potassium Iodide Solution

Dissolve 15g KI in 100ml distilled water

D. Lead Acetate Solution

Dissolve 10g lead acetate in 100 ml of distilled water

E. Stannous Chloride

Dissolve 40 g $SnCl_2 . 2H_2O$in 100 ml Concentrated HCl

F. Silver Diethyl Dithiocarbamate Solution

Dissolve it in 200 ml pyridine.

G. Standard Arsenic Solution

Dissolve 1.32 mg of arsenic trioxide and 4g NaOH in 10 ml distilled water and make volume to 1 liter. Dilute this solution 100 times (10-100ml) it will contains 10mg As/ml. This is standard solution.

Procedure

1. Take 50ml of sample in arsenic estimation apparatus and add 5 ml concentrated HCl, 2ml KI and 0.5ml (8-9drops) stannous chloride solution.
2. Mix the solution and allow it to stand for 15 minutes, meanwhile put the glasswool in the scrubber soaked with lead acetate solution.
3. In an absorber tube, add 4.0ml silver diethyl dithio carbamate solution.
4. In generator, add 3g Zn and connect the assembly as soon as possible.
5. Slightly heat the generator for complete evolution of arsine.
6. Allow it to stand for 30 min.
7. Measure the absorbance of the solution under spectrophotometer at 535 nm.
8. Prepare standard curve with standard solution from 0.0-10ml diluted to 50 ml follows same steps as for sample.

Calculation

$$As\ mg/l = \frac{\mu g\ As}{ml\ of\ sample\ taken}$$

Flow Diagram 5.5: Arsenic Measurement

Take 50ml of sample in arsenic estimation apparatus & add 5 ml conc. HCl, 2ml KI & 0.5ml (8-9 drops) stannous chloride solution

↓

Mix the solution & allow it to stand for 15 min, meanwhile put the glasswool in the scrubber soaked with lead acetate solution

↓

In an absorber tube, add 4.0ml silver diethyl dithio carbamate solution

↓

In generator, add 3g Zn & connect the assembly as soon as possible

↓

Slightly heat the generator for complete evolution of arsine

↓

Allow it to stand for 30 min.

↓

Measure the absorbance of the solution under spectrophotometer at 535 nm.

Sheet 5.3: Data Sheet for Estimation of Arsenic

Sl.No.	Site/Source	Reading						Average	Standard Deviation
		A	B	C	D	E	F		
1.									
2.									
3.									
4.									
5.									
6.									
7.									
8.									
9.									
10.									

(D) Manganese

Principle

In an acidic solution manganese compounds are oxidized to permangnate by using persulphate. The resulting colour will be measured under spectrophotometer at 545 nm.

Apparatus

1. Beaker
2. Spectrophotometer
3. Hot Plate
4. Measuring Cylinder.

Reagents

1. Standard Manganese Solution

Dissolve 2gm manganese metal in 100 ml H_2SO_4 (7%) and make volume to 1 liter using distilled water. Dilute it 40 times *i.e.* (25ml-100ml). This is standard manganese solution. (1ml =50mg Mn).

2. Special Reagent

Dissolve 75g mercuric sulphate in previously mixed 400 ml concentrated H_2SO_4 and 200ml distilled water. Add 200ml 85% H_3PO_4 (phosphoric acid) and 35mg Ag NO_3 and finally make volume to 1 liter with distilled water.

3. Ammonium Persulphate

Solid.

Procedure

1. Take 100ml of sample in a beaker.
2. Add 5ml of special reagent and heat beaker over hot plate.
3. When sample remains less than half (about 45ml) add 1ml ammonium persulphate while boiling on hot plate.
4. After one minute cool the beakers under tap water.
5. Measure the absorbance at 545 nm using spectrophotometer.
6. Prepare the standard curve using suitable dilutions as taken for sample determination.

Note: For sample having turbidity suspended matter (effluent of industries) use HNO_3-H_2SO_4 digestion as described in chromium and after that add 5 ml special reagent and follow steps as given in manganese procedure.

Flow Diagram 5.6: Manganese Measurement

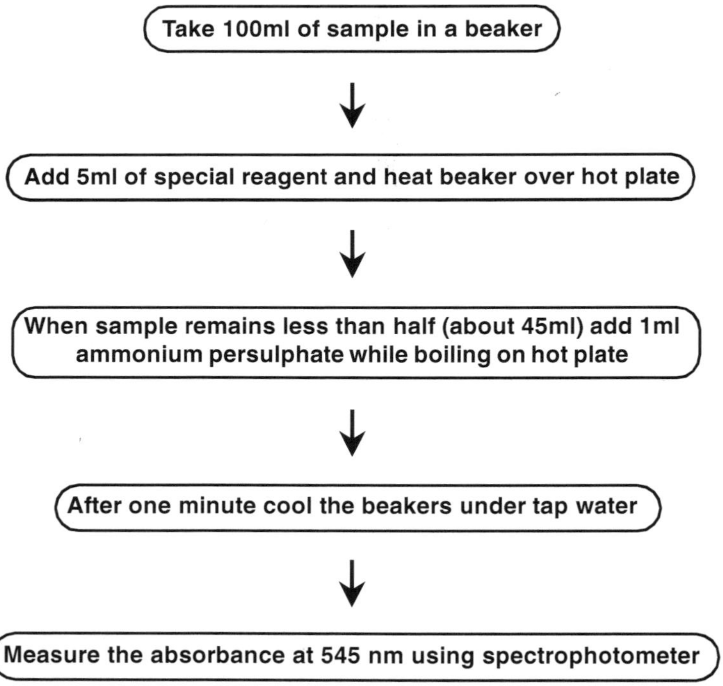

Sheet: 5.4: Data Sheet for Estimation of Manganese

Sl.No.	Site/Source	Reading						Average	Standard Deviation
		A	B	C	D	E	F		
1.									
2.									
3.									
4.									
5.									
6.									
7.									
8.									
9.									
10.									

(E) Lead

Principle

In an alkaline condition (11.5 pH) lead react with dithiozone and forms lead dithizonate. This is soluble in chloroform in presence of alkaline cyanide. The green coloured dithizone is not extracted by chloroform.

Apparatus

1. Separating funnel
2. Spectrophotometer.

Regents

1. Standard Lead Solution

Dissolve 1.599gm of lead nitrate in distilled water-add 10ml concentrated HNO_3 and make volume to 1 liter with distilled water (1ml-.mg Pb). Dilute this solution to 100 times (10ml-100ml). This is standard lead solution (1ml-1mg Pb).

2. Dithizone Working Solution

Dissolve 1.0g dithizone in 100 ml $CHCO_3$. Take 12ml of this solution in separating funnel. Add 20ml 0.5 N ammonia solution and shake it. Filter aqueous layer through filter paper. This is dithizone working solution.

3. Ammonia Solution

Dilute 3.5ml ammonia solution in 100ml distilled water.

4. Sodium Hexameta Phosphate Solution

Dissolve 10gm sodium hexameta phosphate (lead free) in 100 ml distilled water.

5. Hydroxyl Amine Hydrochloride Solution

Dissolve 1gm hydroxyl amine hydrochloride in 100 ml distilled water.

6. Alkaline Cyanide Solution

Dissolve 3g sodium sulphite in previously mixed 340ml ammonia and 680 ml distilled water and add 30 ml 1% potassium cyanide and mix well.

Procedure

1. Take 50ml of sample in separating funnel.
2. Add 1ml sodium hexameta phosphate, 1ml hydroxyl amine hydrochloride, 30 ml alkaline cyanide, 0.5ml dithizone working solution and 10 ml chloroform.
3. Shake the funnel and allow it to stand for separation of layers.
4. Take the chloroform layer in optical cell and measure absorbance at 510nm using spectrophotometer with green filters.
5. Prepare a standard curve using various standard dilutions and follow the same steps as for sample determination.

Note: If sample contain colours and organic matters, follow HNO_3-H_2SO_4 digestion before starting process.

Flow Diagram 5.7: Lead Measurement

Take 50ml of sample in separating funnel

↓

Add 1ml sodium hexameta phosphate, 1ml hydroxyl amine hydrochloride, 30 ml alkaline cyanide 0.5ml dithizone working soln. & 10 ml chloroform

↓

Shake the funnel & allow it to stand for separation of layers

↓

Take the chloroform layer in optical cell & measure absorbance at 510nm using spectrophotometer with green filters

Sheet 5.5: Data Sheet for Estimation of Lead

Sl.No.	*Site/Source*	*Reading*						*Average*	*Standard Deviation*
		A	*B*	*C*	*D*	*E*	*F*		
1.									
2.									
3.									
4.									
5.									
6.									
7.									
8.									
9.									
10.									

(F) Zinc

Principle

In an alkaline condition when zincon (2-carboxy-2-hydroxy-5-sulphoformazyl benzene) is added to sample containing zinc a blue coloured complex is formed which can be measured under spectrophotometer.

Apparatus

1. Conical flask
2. Measuring cylinder
3. Spectrophotometer.

Reagents

1. Standard Zinc Solution

Dissolve 100mg of Zn in 100ml of (1+1) HCl and finally make volume to 1 liter (1ml=100 m gZn).Now dilute this solution to 10 times (10-100ml). This is standard Zn solution. (1ml=10m g Zn).

2. Sodium Ascorbate

Granular.

3. Potassium Cyanide Solution

Dissolve 1g KCN in 100ml distilled water.

4. HCl

Concentrated.

5. Sodium Hydroxide (6N)

Dissolve 240g of NaOH pellets in 1liter distilled water.

6. Zincon Reagent

Dissolve 130 mg zincon powder is 100 ml methanol.

7. Chloral Hydrate Solution

Dissolve 10g chloral hydrate in 100 ml distilled water.

8. Buffer Solution

Dilute 213ml of 1N NaOH in distilled water. Add 37.8 g KCl and 31 g H_3BO_3 and finally make volume to 1 liter using distilled water.

Procedure

1. Take 10 ml of sample in a conical flask.
2. Add 0.5 g sodium ascorbate, 1ml KCN, 5ml buffer solution and 3ml Zincon solution to the sample.
3. Mix thoroughly and add 3ml chloral hydrate.
4. Measure absorbance using spectrophotometer at 620nm after 5-7 minutes.
5. Prepare standard curve with standard solution in range of 0.0ml-70ml diluted to 70ml with distilled water following same steps as taken for sample determination.

Note: If total Zn is to be determined, add 1ml HCl in 50ml sample and adjust its pH to 7 before addition of any solution.

Calculation

$$Zn\ mg/l = \frac{\mu g\ Zn}{ml\ of\ sample\ taken}$$

Flow Diagram 5.8: Zinc Measurement

Take 10 ml of sample in a conical flask

↓

Add 0.5 g sodium ascorbate, 1ml KCN, 5ml buffer soln. & 3 ml Zincon solution to the sample

↓

Mix thoroughly and add 3 ml chloral hydrate

↓

Measure absorbance using spectrophotometer at 620nm after 5-7 minutes

Sheet 5.6: Data Sheet for Estimation of Zinc

Sl.No.	*Site/Source*	*Reading*						*Average*	*Standard Deviation*
		A	*B*	*C*	*D*	*E*	*F*		
1.									
2.									
3.									
4.									
5.									
6.									
7.									
8.									
9.									
10.									

(G) Iron

Principle

When hydrochloric acid and hydroxylamine is added to sample containing iron it converts to its ferrous state which with addition of 1, 10 phenonthroline give a complex of orange red colour which would be examined spectrophotometrically to know concentration of iron.

Apparatus

1. Conical flask
2. Measuring cylinder.

Reagent

1. HCl

Concentrated.

2. Hydroxyl Amine Hydrochloride

Dissolve 10gm of Hydroxyl amine hydrochloride in 100 ml distilled water.

3. Ammonium Acetate Solution

Dissolve 250 gm ammonium acetate in 150 ml distilled water and 700 ml glacial acetic acid.

4. Phenonthroline Solution

Dissolve 100mg 1,10 phenonthroline monohydrate in 100ml distiled water by heating.

5. Standard Iron Solution

Dissolve 1.4g ferrous ammonium sulphate in 20ml concentrated H_2SO_4 and 50 ml distilled water, add 0.1N $KMnO_4$ and dilute it to 1 liter. (200 mg Fe/lt). Dilute it 20 times (50-100 ml). This is standard iron solution (10mg fe/lt).

Procedure

1. Take 50ml of sample in a conical flask.
2. Add 2ml concentrated, HCl & 1ml hydroxylamine hydrochloride solution & boil it.
3. Cool when sample remain half of it quantity and add 10ml ammonium acetate and 2ml phenonthroline solution. (add more if sample have interferences).
4. An orange red colour will appear.
5. Make the final volume of sample to 100 ml and measure absorbance at 510nm using spectrophotometer exactly after 10 minutes.
6. Prepare a standard curve using various dilutions (0-5 ml) of standard solution following same steps as for sample determination.

Flow Diagram 5.9: Iron Measurement

Take 50ml of sample in a conical flask

↓

Add 2ml conc, HCl & 1ml hydroxylamine hydrochloride solution & boil it

↓

Cool when sample remain half of it quantity & add 10ml amm. Acetate & 2ml phenonthroline solution (add more if sample have interferences)

↓

An orange red colour will appear

↓

Make the final volume of sample to 100 ml & measure absorbance at 510nm using spectrophotometer exactly after 10 minutes

Sheet 5.7: Data Sheet for Estimation of Iron

Sl.No.	*Site/Source*	*Reading*						*Average*	*Standard Deviation*
		A	*B*	*C*	*D*	*E*	*F*		
1.									
2.									
3.									
4.									
5.									
6.									
7.									
8.									
9.									
10.									

Note: It should be recommended that all heavy metals should be measured under Atomic Absorption spectrophotometer at various wave lengths.

Chapter 6
Biological Analysis of Water

(I) Freshwater Organisms

On the Basis of their Mode of Life

On the basis of their mode of life the fresh water organisms can be classified into following categories:

(*a*) Plankton
(*b*) Periphytons
(*c*) Neustons
(*d*) Nektons
(*e*) Psammons
(*f*) Pleuston
(*g*) Benthos

A. Plankton

The word Plankton is derived from Greek word ' Plankton' meaning free swimming. These are the organisms which keep on floating on the surface of water and are always at the mercy of water currents or these are microscopic communities of plants (phytoplankton) or animals (Zoo plankton) found suspended in natural waters. These have usually little or no résistance to water currents and are found free floating. They are further classified as follows:

(1) On the Basis of their Life Cycle

Holoplankton

These plankton remain plankton throughout their life.

Meroplankton

These organisms spend only a part of their life cycle as planktonic organisms, *e.g.* Larvae of fishes, Prawns etc.

(2) On the Basis of their Habit

Phytoplankton

These are the organisms which have the capacity to synthesize food, *e.g. Navicula, Volvox, Oscillatoria* species *etc.*

Zooplankton

These organisms donot have the capacity to synthesize food. They depend on phytoplankton for their food, *e.g.* Rotifers, Cladocerans etc.

(3) On the Basis of their External Environment or Habitat

Limnoplankton

Plankton inhabiting the lakes are called limnoplankton.

Rheoplankton or Potamoplankton

Plankton inhabiting in running water are called Rheoplankton or Potamoplankton.

Heleoplankton

Plankton inhabiting the ponds are called Heleoplankton.

Haliplankton

Plankton inhabiting the sands are called as Haliplankton.

Hypalinyroplankton

Plankton inhabiting brackish water are called Hypalinyroplankton.

(4) On the Basis of their Size

Megaplankton

1 cm.

Mesoplankton

1 mm-1 cm.

Macroplankton

0.5 mm-1 mm.

Microplankton

0.06 mm- 0.5 mm.

Nannoplankton

0.005 mm-0.06 mm.

Ultraplankton

0.0005 mm-0.005 mm.

(5) On the Basis of Phytoplanktonic Content

Euplankton

These are the true plankton.

Pseudoplankton

These have debris mixed with plankton.

(6) On the Basis of Filterability

Net Retainable

These can be retained by net during filteration.

Nannop

These are not retained by the net.

(7) On the Basis of their Origin

Autochthonous

These arise from within the water body.

Allochthonous

These are introduced from outside into the water body.

Collection of Samples (Phytoplankton and Zooplankton)

As the plankton present in water varies in sizes therefore to collect plankton, different types of samplers such as Vandorn etc is used. For collection of plankton of bigger type most common sampler i.e. plankton net is used.

Most commonly used plankton net is conical in shape with ring of stainless steel with a collection tube at the bottom. The design of net is shown in Fig-6.1. The use of plankton net involves the mesh sizes, type of material used, amount of water filtered. Nylon monofilament mesh is preferred because of its mesh size accuracy and durability. The known amount of water was filtered and concentrated in the tube present at lower end of net.

The sample collected in the tube is to be observed under microscope immediately after collection. If in case the sampling site is far from laboratory the sample must be preserved using suitable preservative.

Preservation of Sample

The most suitable phytoplankton preservative is Lugol's solution. Instead of Lugol's solution, we should also preserve the sample using 4% formaline. Add 0.5ml of Lugol's solution to per 100ml of sample collected, Lugol's solution should preserve cilia and flagella of plankton and also stains them which would help in identifying them. It is also suggested that Lugol's solution is not good for long time storage because it would results in shrinkage of plankton.

Preparation of Lugol's Solution

Dissolve 100gm of KI and 5gm iodine crystals in 20ml of distilled water, then add 50 ml distilled water in which 5gm anhydrous sodium acetate has been dissolved.

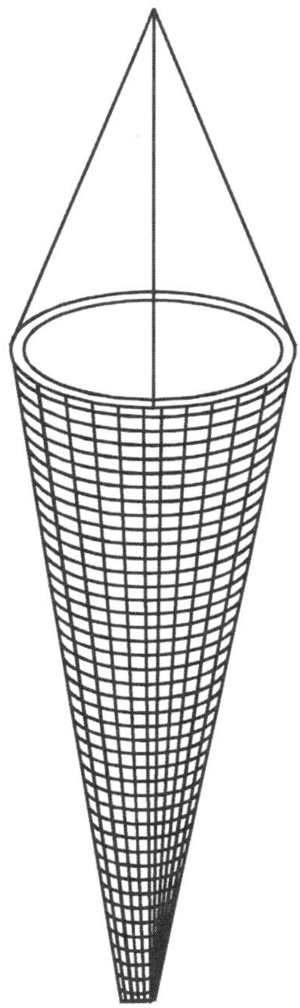

Figure 6.1: Simple Plankton Net

Qualitative and Quantitative Analysis of Plankton (Both)

Device used for this analysis is Sedgwick Rafter counting cell. Sedgwick Rafter cell is approximately 50mm long, 20mm wide and 1 mm deep. The total volume of the cell is 1ml. The diagram of sedgwick rafter counting cell is shown in Figure 6.2.

Shake the sample well in the sample tube so that the organisms may distribute, now transfer the sample using large bore pipette of 1ml. Place the cover slip on the cell avoiding any kind of air bubble. Before counting wait for 10-15 minutes so as all plankton may settle down.

Now place the cell under microscope and count all the plankton present in cell by moving it horizontally and vertically. Repeat the process several times. Counting of organisms will be done by applying the following formula.

$$n = \frac{(a \times 1000)\, c}{l}$$

where,

n: Number of plankton / liter of water

a: Average no. of plankton in one small counting chamber of S-R cell.

c: ml of plankton concentrate

l: Volume of original water filtered in liter.

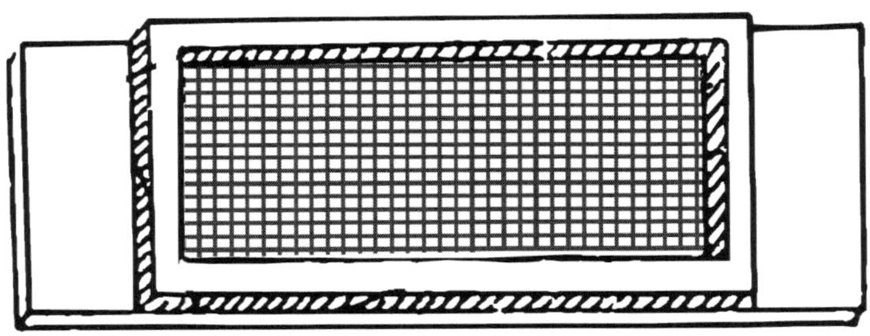

Figure 6.2: Sedgwick Rafter Counting Cell

Flow Diagram 6.1: Plankton Measurement

Filter known amount of water sample from plankton net

↓

Remove collection tube from bottom of net

↓

Add suitable preservative in the collected sample

↓

Count the number under microscope using Sedgwick Rafter counting cell

↓

Apply the observed number of plankton in formula

Sheet 6.1: Data Sheet for Estimation of Plankton

Sl.No.	*Site/Source*	*Reading*						*Average*	*Standard Deviation*
		A	*B*	*C*	*D*	*E*	*F*		
1.									
2.									
3.									
4.									
5.									
6.									
7.									
8.									
9.									
10.									

B. Periphyton

These are the aquatic organisms which remain attached to the submerged surfaces of the various objects present in water. These are further of four types:

(*i*) Epiphytic

These are the organisms which attach themselves to the submerged parts of stems and leaves of aquatic plants.

(*ii*) Epizoic

These are the organisms which attach themselves to exposed surfaces of submerged animals like shells of molluscs etc.

(*iii*) Epilithic

These organisms attach themselves on the surface of submerged rocks and stones.

(*iv*) Epipelic

These organisms attach themselves with the sediments at the bottom of water body.

C. Neuston

These are the organisms which are associated with the surface film of water. These organisms occur at the interface between air and water.

They are of further two types:

(*i*) Supraneustori or Epineustori

These are the organisms which are associated with the dorsal surface or upper surface of the surface film of water, *e.g.* Hydrometra among animals and Azolla among plants.

(*ii*) Infraneustori or Hyponeustori

These are the organisms which are associated with the lower surface of the surface film of water, *e.g.* Insect larvae of mosquitoes.

D. Nektons

These are the free swimming organisms inside the water body. These have various adaptations for living inside the water body. These include fishes, crabs, prawns etc.

E. Psammons

These are the organisms which are found in the interstidial waters. These occur in the intertidal zone of the beaches, on the sides of lake etc.

These show a vertical stratification in water. The autotrophs are placed in the upper lighted zone. The consumers are present in the lower portions. The anaerobic bacteria are only present in the anaerobic regions. The vegetation in this zone is quiet different from the vegetation occurring in other areas of the water body.

F. Pleuston

These are the rooted free floating organisms in water body, *e.g. Azolla, Anabena* etc.

G. Benthos

These are relatively large sized organisms and are present on the substratum of lakes, pond, rivers, streams etc. The major macro invertebrates found in fresh water are annelida, molluscs, worms and insects etc. These are of three types:

(*i*) Phytobenthos

These are the producer organisms which remain attached with the substratum with the help of roots. These include the macrophytes which are further classified as:-

(*a*) *Emergent Plants*

These are the plants which have a part of their stem and roots submerged in water while part of stem, leaves and flowers are always outside water.

(*b*) *Floating Leaf Plants*

These macrophytes have their roots embedded in the substratum but their leaves float on water. The leaves have long petioles, *e.g. Brassica species, Nelumbium species* etc.

(*c*) *Floating Submerged Plants*

These macrophytes have roots embedded in substratum and they keep on floating inside the water body, *e.g. Potamageton* species, *Hydrilla* species etc.

(*ii*) Zoobenthos

These are the aquatic animals which remain attached to the substratum. These form an important link in the decomposer food chain. These organisms breakdown the complex molecules into simpler forms and thus help in nutrient recyclying. These are of two types:

(*a*) *Epifauna*

These are benthic forms which either move about freely or are attached.

(*b*) *Infauna*

These are benthic fauna which dig and live in burrows.

Collection of Sample

Various types of device are used to collect Benthos sample but the most common and easiest method is by using drag net, insect net or by using forceps.

Preservation of Sample

The collected sample is washed away from the substratum or sediments. These can be preserved by using 10% formaline solution.

Identification

Examine organism visually or under microscope and match the organisms with Ward and Whipple (1992).

Flow Diagram 6.2: Benthos Measurement

Collect the sample using suitable sampling device

↓

Wash sample carefully

↓

Preserve the sample using suitable preservative

↓

Examine the sample visually or under microscope

Sheet 6.2: Data Sheet for Estimation of Benthos

Sl.No.	*Site/Source*	*Reading*						*Average*	*Standard Deviation*
		A	*B*	*C*	*D*	*E*	*F*		
1.									
2.									
3.									
4.									
5.									
6.									
7.									
8.									
9.									
10.									

(II) Fishes

Fishes are the major components of most aquatic system.

Collection of Fishes

Fishes can be collected from rivers, lakes, ponds, streams etc. by the following methods:

1. Jaal
2. Fatiyala (triangular net)
3. Pot
4. Scoop nets
5. Fine cloth
6. Foams
7. Angling.

Preservation of Fishes

Fishes should be preserved in 5% formaline solution.

Identification

Fishes can be identified according to Day (1978), Shrivastava (1968), Badola (1979), Jhingran (1975), Sehgal (1978) and Lindberg (1976).

Flow Diagram 6.3: Fish Collection and Identification

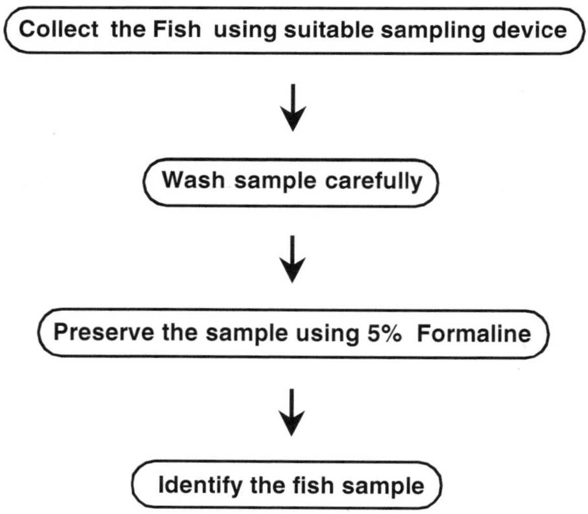

Sheet 6.3: Data Sheet for Fish Identification and Collection

Present Scientific Name	*Name According to Day*	*Name According to Hamilton*	*Local Name*
	Phylum		
	Subphylum		
	Superclass		
	Series		
	Class		
	Subclass		
	Order		
	Suborder		
	Family		
	Sub Family		
	Genus		

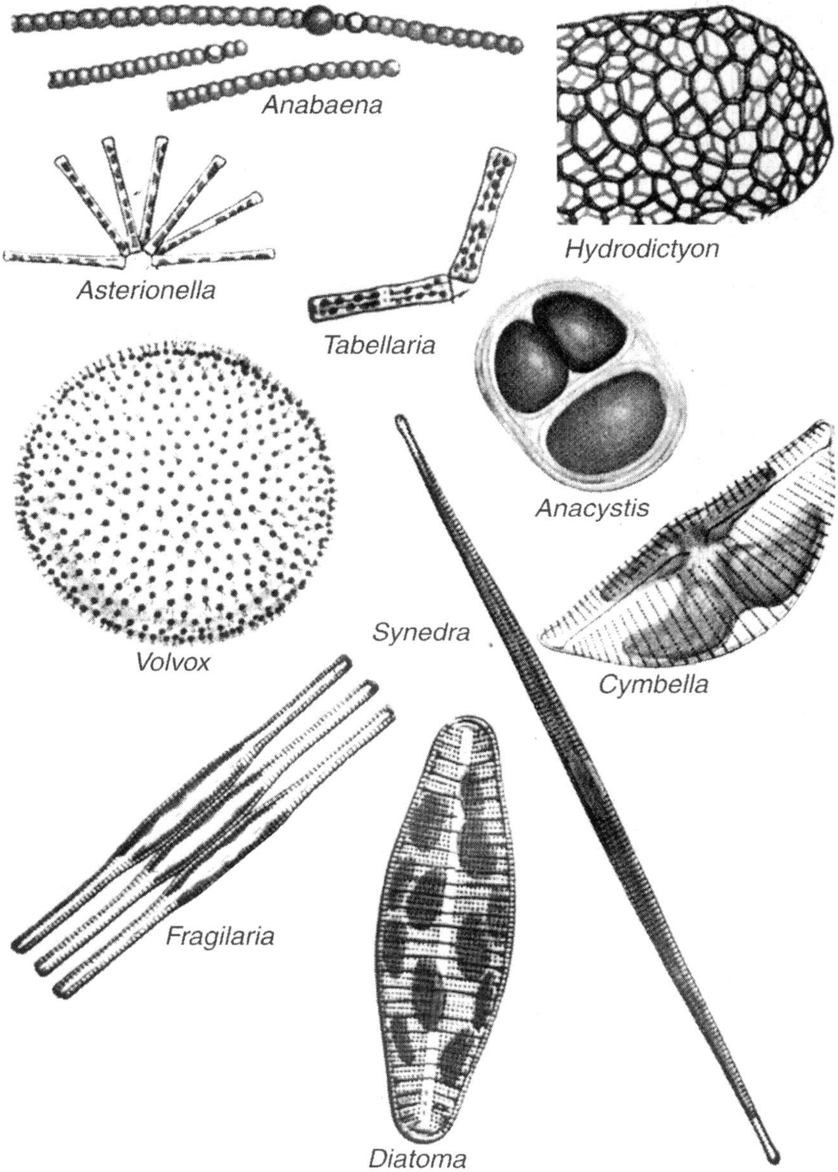

Plate 6.1: Plankton

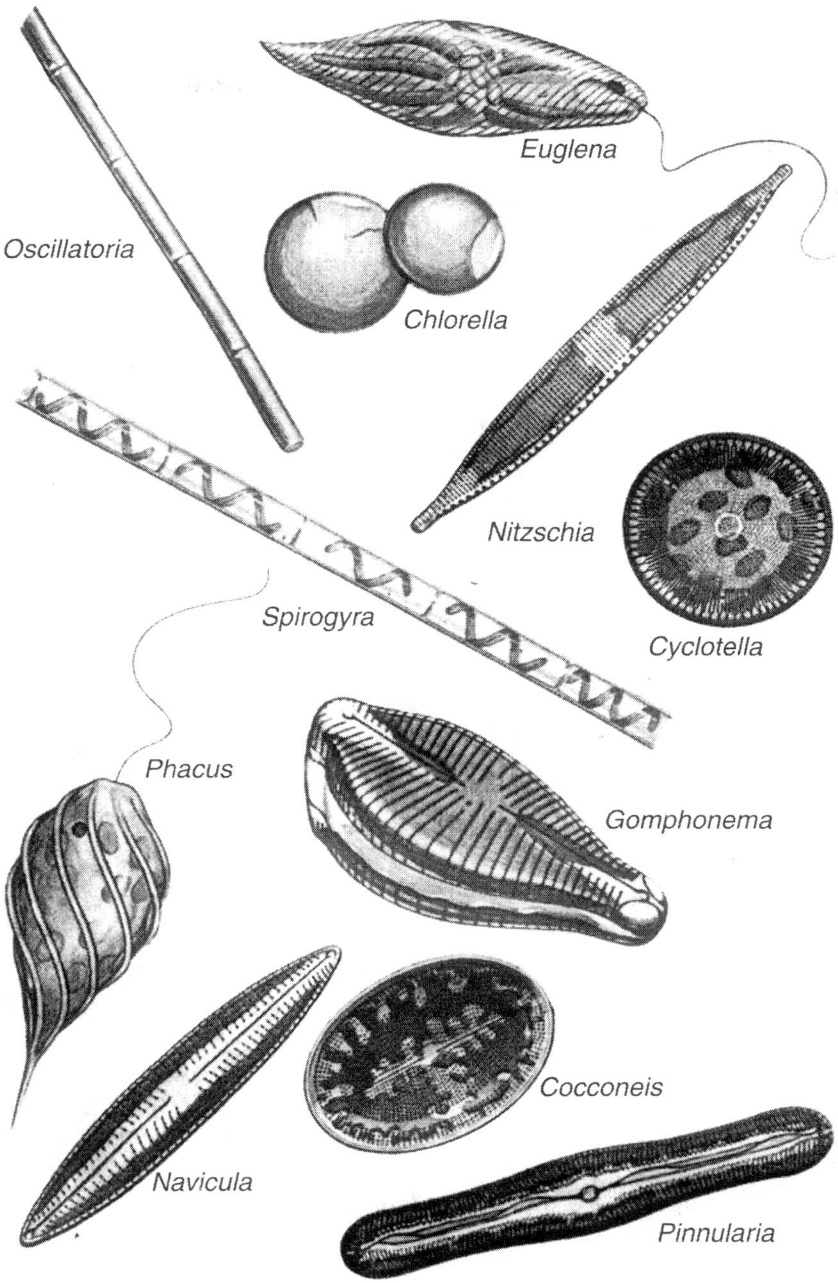

Plate 6.2: Plankton

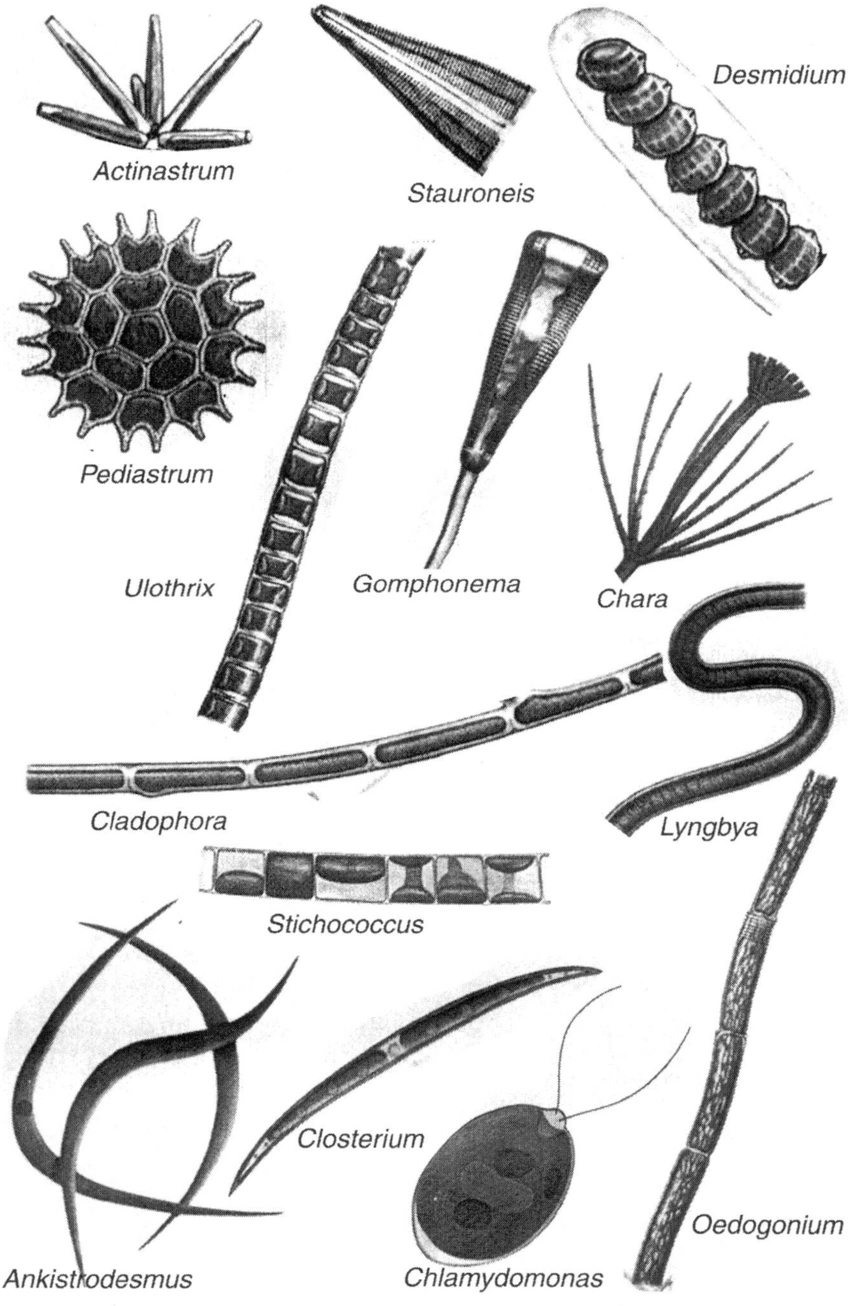

Plate 6.3: Plankton

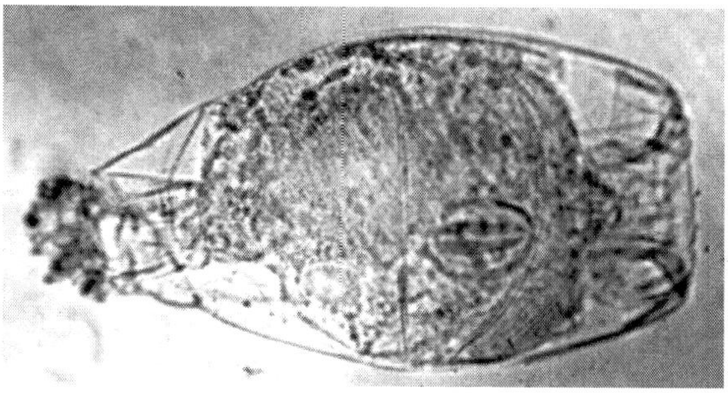

Vonoyella globosa

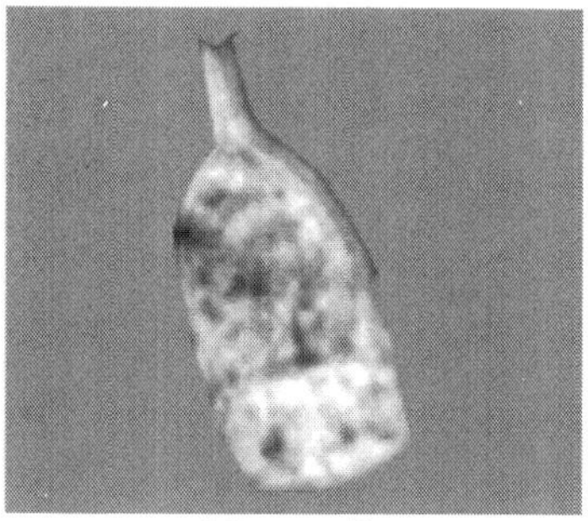

Keratella

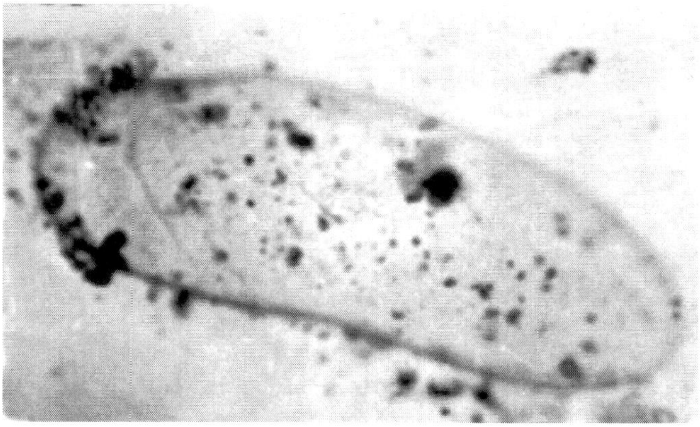

Cypris Subglobosa

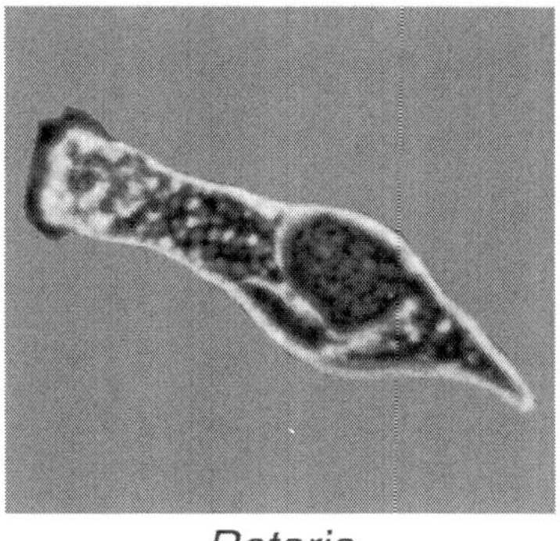

Rotaria

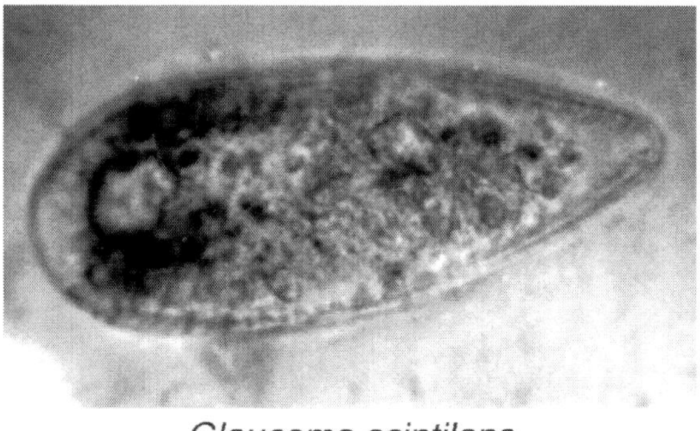

Glaucoma scintilans

Plate 6.4: Zooplankton

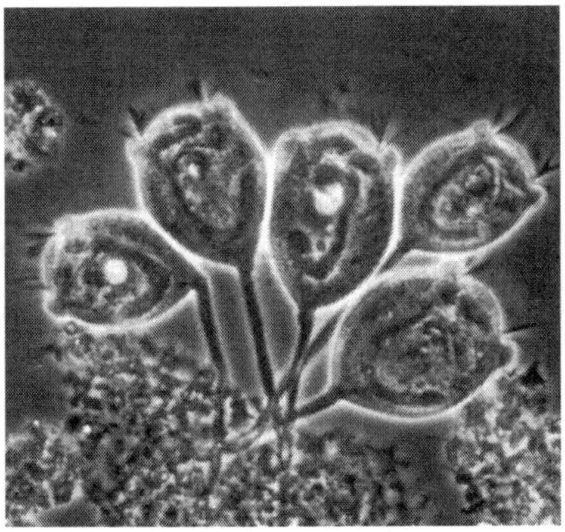

Vorticella

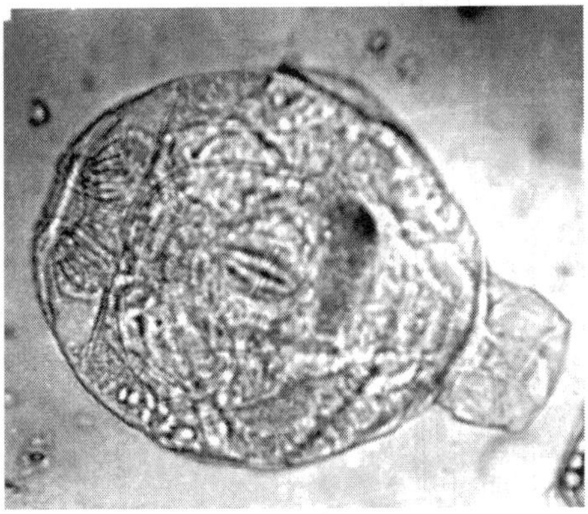

Bdelloid

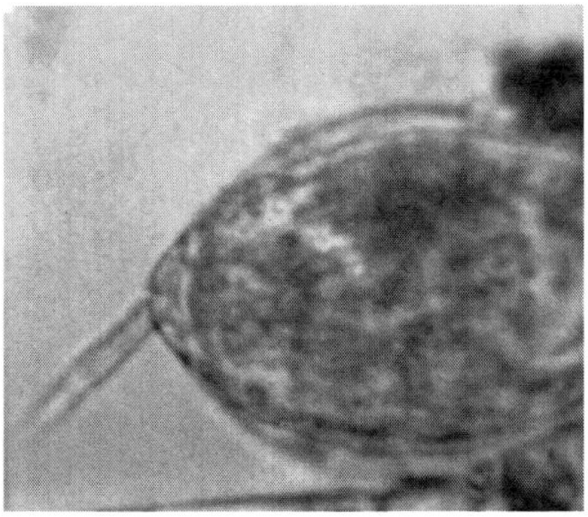

Monostyla lunaris

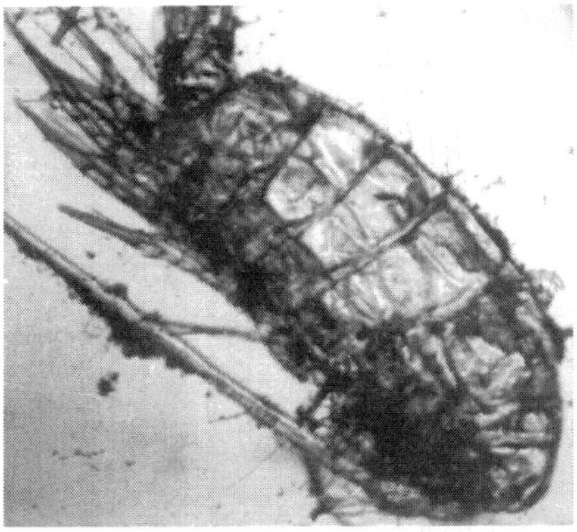

Diaptomus

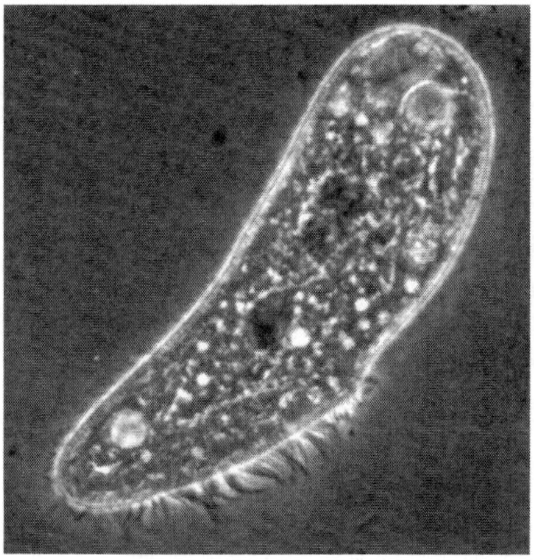

Blepharisma

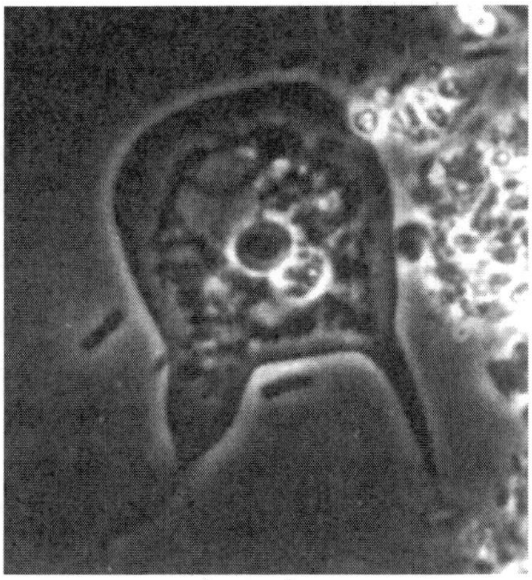

Amoeba

Plate 6.5: Zooplankton

Some Freshwater Fishes

Mastacembelus armatus

Classification

Series	:	**Pisces**
Class	:	**Teleostomi**
Subclass	:	**Actinopterygii**
Order	:	**Mastacembeliformes**
Family	:	**Mastacembelidae**
Genus	:	***Mastacembelus***
Species	:	***armatus***

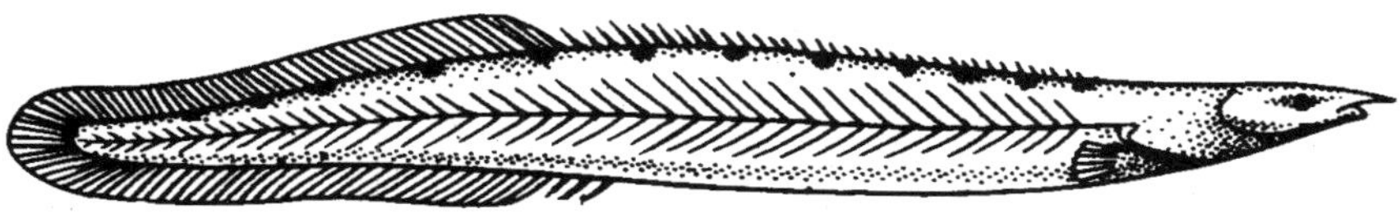

Fin Formula

D. 37-39/78-82, P. 23-25, A . 3/75-78.

Comments

(a) Length of head 5.75 to 6.5 and the height of the body is 10.3 to 11.25 in the total length (15.2–21 cm).

(b) *Eyes*: Diameter 11 mm to 11.6 mm in the length of head, 4 mm diameter from the end of snout and 0.8 to 1 in the interorbital width.

(c) *Fins*: The dorsal spines commence over the middle of the pectoral fin and increase in size from anterior to posterior, the last but one is the longest.. Ventral fin is absent.

(d) *Scales*: Very minute with 30 rows between the lateral line and the base of the first dorsal ray.

(e) *Colour*: Rich brown becoming lighter on the abdomen.

(f) *Distribution*: Baluchistan, Pakistan, India, Ceylon, Nepal, Burma,Malaya, S. China, Sumatra and Java.

Anabas testudineus

Classification

Series	:	**Pisces**
Class	:	**Telesotomi**
Subclass	:	**Actinopterygii**
Order	:	**Perciformes**
Sub order	:	**Anabantaidae**
Family	:	**Anabantidae**
Genus	:	***Anabas***
Species	:	***testudineus***

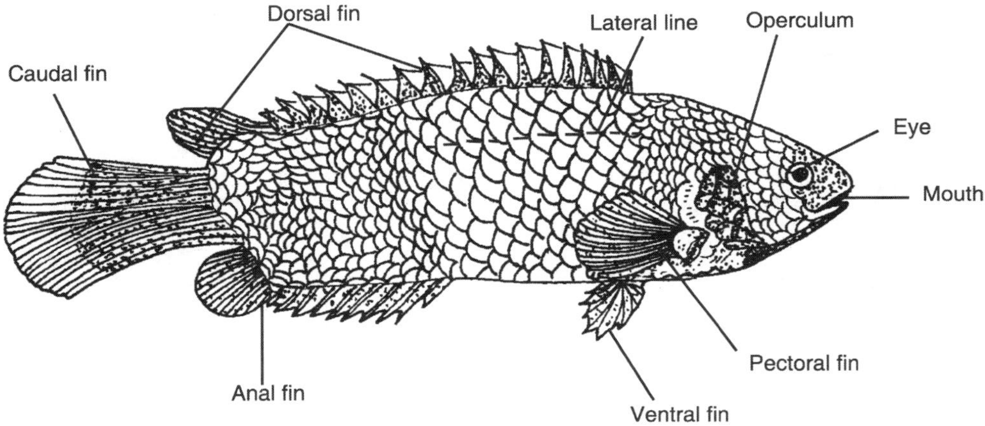

Fin Formula

D. 17/10, P.15, V.1/5, A.11, C.17, L.I.29, L.tr. 4/9

Comments

(*a*) Length of head 3.6 of caudal 5.4 and the height of the body is 3.6 in the total length (13 to 13.8 cm).

(*b*) *Eyes*: Diameter 5 mm in the length of the head and from the end of snout and 7.5 in the interorbital width preorbitals is strongly denticulated but the posterior edge of the opercle, sub opercle and interopercle are strongly spinous.

(*c*) *Teeth*: The outer row of villi from teeth in the jaws are the largest vomers have smaller teeth while the palatine are tooth less.

(*d*) *Fins*: Spinous portion of the dorsal fin is 8 in entire length of the fin while the soft portion is higher than spinous one.

(*e*) *Scales*: Strongly ctenoid some scales are found over the base of the dorsal, pectoral and anal and caudal fins also.

(*f*) *Lateral Line*: It is intrupped at 18th scale, 10th scale is a perforated one and is formed just below 18th one.

(*g*) *Distribution*: Fresh waters and estuaries of oriental regions.

Local Name: Kawai and Sumha

Chanda nama

Classification

Series	:	**Pisces**
Class	:	**Teleostomii**
Subclass	:	**Actinopterygii**
Order	:	**Perciformes**
Sub order	:	**Percoidei**
Family	:	**Centropomidae**
Genus	:	***Chanda***
Species	:	***nama***

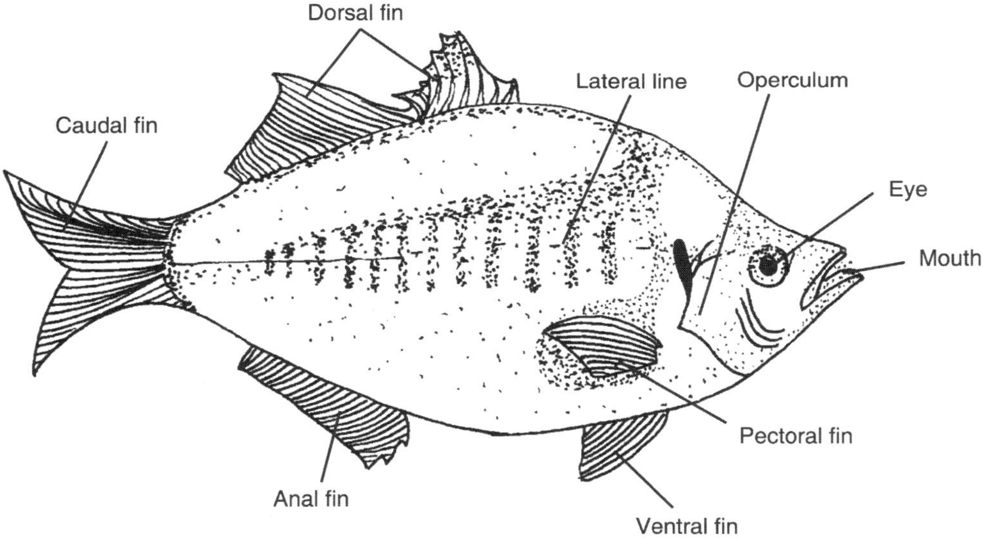

Fin Formula

D.1+7/1/16, P.12, V.1/5, A.3/16, C.17, Lr.-scales deciduous.

Comments

(*a*) Length of head 4.05 (4-4.2); of caudal 4.86(4.75-5) and height of body varies with age and is 4.62 (4.5-4.75) in youngs and 3.45 (3.3-3.5) in older one in total length (3.4 to 3.8 cm).

(*b*) *Eyes*: Diameter 3 mm in the length of head, 0.75 from the end of snout and 0.75 in the interorbital width, lower jaw much longer than upper.

(*c*) *Teeth*: Two large curved canines on either side of lower jaw present. Both jaws with outer and inner row of teeth.

(*d*) *Fins*: The 3rd dorsal spine is longest, rays gradually decreases in length. 3rd spine largest, strongest as in dorsal. Caudal fin deeply forked.

(*e*) *Scales*: Minute.

(*f*) *Lateral line*: Broken or absent.

(*g*) *Colour*: Fresh whitish yellow with black dots over body. Summit of head and top of eyes are black. Anal fin has spots at the base of spine.

(*h*) *Distribution*: Throughout the fresh waters of India, Pakistan, Burma

(*i*) Key characteristic; lower jaw with 2 pairs of large caniniform teeth.

Channa striatus

Classification

Series	:	**Pisces**
Class	:	**Teleostomi**
Subclass	:	**Actinopterygii**
Order	:	**Ophiocephaliformes**
Family	:	**Ophiocephalidae**
Genus	:	***Channa***
Species	:	***striatus***

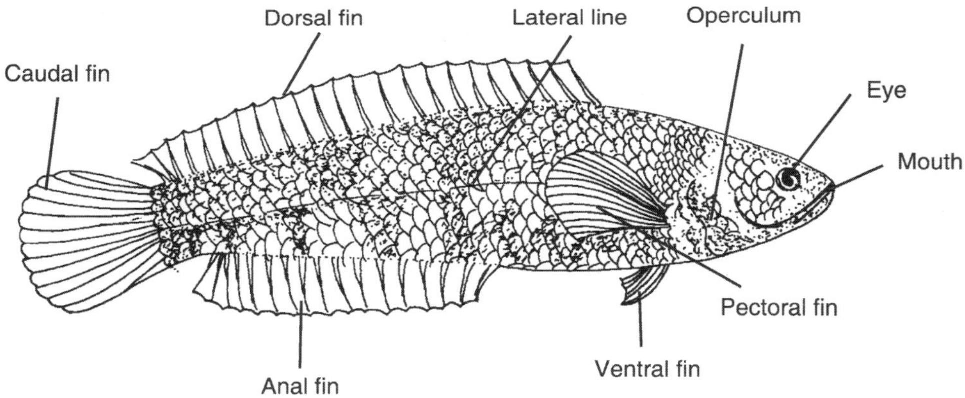

Fin Formula

B.V, D.37-45, P.17, V.6, A.23-26, C.13, L.I.50-57, L.tr. 4½-7/9-7or 5½-8/10-9

Comments

(*a*) Length of head 3.75 of Caudal fin 6 and height of body 7.6 in total length (28.5 cm).

(*b*) *Eyes*: Diameter 7.15 mm in the length of head, 1 from the end of snout and 1.56 in the interorbital width.

(*c*) *Teeth*: In lower jaw with an inner conical row, palatine with cardiform teeth.

(*d*) *Scales*:Large irregular shaped scales on summit of head, a row of 18-20 predorsal scales. A row of 9 scales between the orbit and the angle of preopercle, 7 row between lateral line and dorsal fin.

(*e*) *Lateral line*: Curves downwards below 12th dorsal ray.

(*f*) *Color*: Dark greyish or blackish dorsally depending upon the locality; yellow, white beneath; cheeks and lower surface of mouth spotted with grey or black.

(*g*) *Distribution*: Throughout the plains of India, Pakistan, Ceylon, Burma, China and Phillipines.

(*h*) *Local name*: Sawr.

Channa gachua

Classification

Series	:	**Pisces**
Class	:	**Teleostomi**
Subclass	:	**Actinopterygii**
Order	:	**Ophiocephaliformes**
Family	:	**Ophiocephalidae**
Genus	:	*Channa*
Species	:	*gachua*

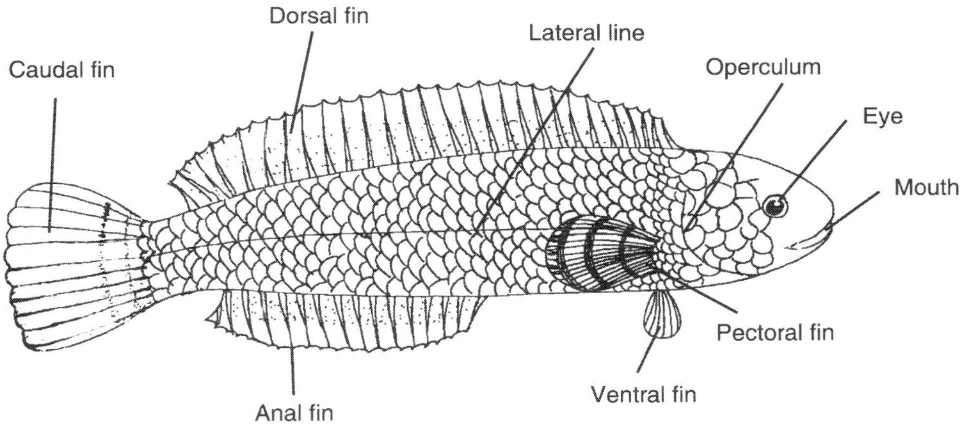

Fin Formula

D.35-37, P.15-16, A.21-23, C.12, L.I.41-45, L.tr. 4/7

Comments

(*a*) Length of head 4.2 (4-4.25); of caudal fin 6.4(6.25-6.5) and height of body 6 in the total length (8 to 13 cm).

(*b*) *Eyes*: Diameter 6 mm is the length of head from the end of snout and 1.5 in the interorbital width.

(c) *Teeth*: In the lower jaw, the inner row of the teeth widely separated and Conical vomer and palatine beer teeth.

(*d*) *Fins*: Pectoral shorter than the length of head behind the eyes. Ventral fin 2/5 of pectoral fin.

(*e*) *Scales*: On the head are broad and irregular. A row of 12 predorsal scales and a row of 4.5 scales between orbit and the angle of the preopercle.

(*f*) *Lateral line*: Curves downwards after 12th scale.

(*g*) *Colour*: Varies according to the habitat. Usually blackish bark, dorsally gradually fainter beneath dorsal, anal, caudal fin state coloured with orange margins. Pectoral fin with a deep blue base, Transversely barred with orange and blue strips, margin being orange coloured. Caudal fin barred body shows white slanting vertical stratification.

(*h*) *Distribution*: Fresh waters of India, Pakistan, Ceylon, Burma and Andamans.

Xenetodon cancila

Classification

Series	:	**Pisces**
Class	:	**Teleostomi**
Subclass	:	**Actinopterygii**
Order	:	**Beloniformes**
Suborder	:	**Scomberescocoidae**
Family	:	**Belonidae**
Genus	:	***Xenetodon***
Species	:	***cancila***

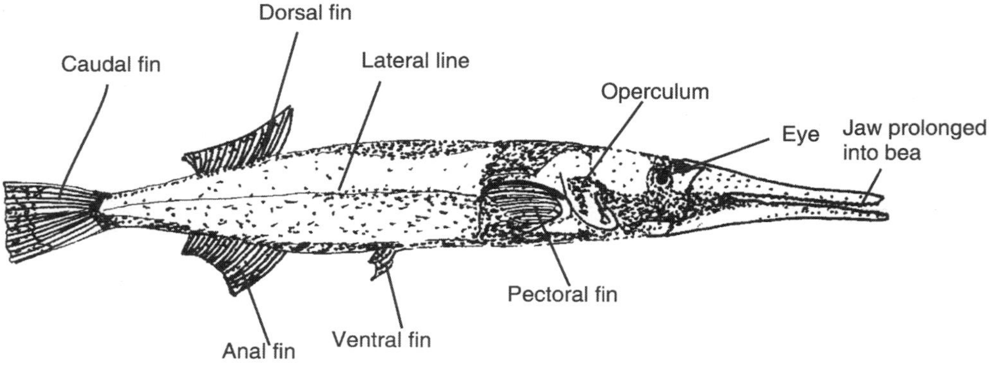

Fin Formula

D. 16-17, p.11, V.6, A.17, c.15

Comments

(*a*) Length of head 2.6 to 2.8 of caudal 10.25 and height of the body is 12 to 12.25 in the total length (21.7 to 24.5 cm).

(*b*) *Eyes*: Diameter 3 mm from the hind edge of opercle and 1in interorbital width.

(c) *Teeth*: Both the jaws are provided with a row of large sharp widely separated teeth and an external row of numerous fine ones which are more on the lower jaw.

(*d*) *Fins*: Dorsal fin commences opposite the anal, caudal fin slightly emarginate.

(*e*) *Scales*: Small and irregular arranged. Some are found on the posterior end of the groove of the head and on the sides of head excepts the opercles.

(*f*) *Colour*: Greenish grey.

(*g*) *Distribution*: Fresh water species found throughout India, extending to Ceylon and Burma.

Heteropneustes fossilis

Classification

Series	:	**Pisces**
Class	:	**Teleostomi**
Subclass	:	**Actinopterygii**
Order	:	**Cypriniformes**
Division	:	**Silun**
Sub order	:	**Siluroidae**
Super family	:	**Siluriodae**
Family	:	**Sacchobranchidae**
Genus	:	***Heteropneustes***
Species	:	***fossilis***

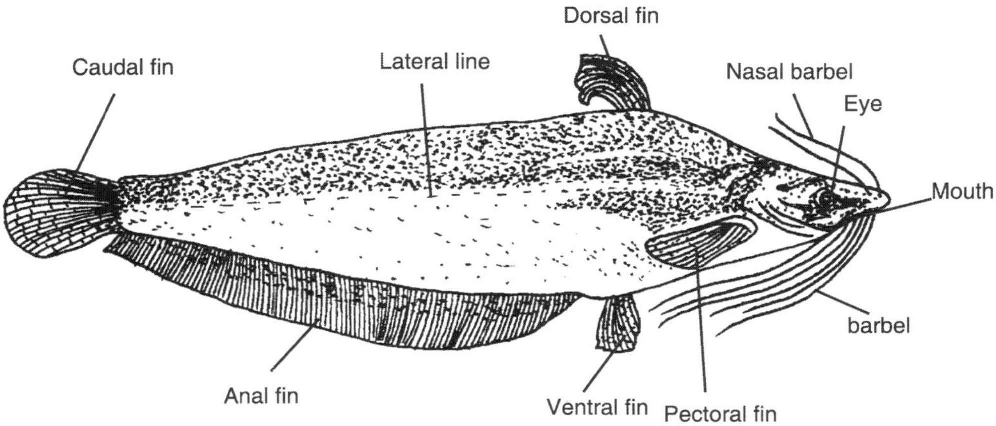

Fin Formula

B.III, D.6-7, P. 1/7, V.6, A.60-79, C.19

Comments

(*a*) Length of head 6.5 to 6.7; of caudal fin 10 in the total length (17.3 to 23.5 cm), and height of body in 6.7-7.

(*b*) *Eyes*: Diameter 8 to 8.75 in the length of head; 2.7 to 3 from the end of snout and 4 to 4.5 in the interorbital width.

(*c*) *Barbels*: 4 pairs, the maxillary extends beyond the pectoral fin.

(*d*) *Fins*: The dorsal fin commence in the anterior 1/3 of body, and the ventral fin reaches up to third or fourth anal ray.

(*e*) *Colour*: Dark leader brown, the youngs are reddish.

(*f*) *Distribution*: Fresh water of India, Pakistan, Ceylon, Burma, and China.

Clupisoma garva

Classification

Series	:	**Pisces**
Class	:	**Teleostomi**
Subclass	:	**Actinopterygii**
Order	:	**Cypriniform**
Division	:	**Siluri**
Suborder	:	**Siluroide**
Super family	:	**Siluroidae**
Famliy	:	**Schibeidae**
Genus	:	***Clupisoma***
Species	:	***garva***

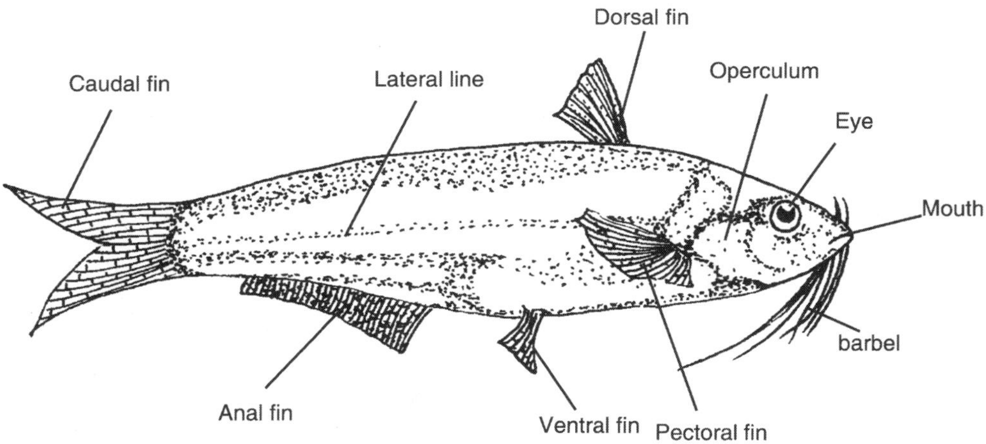

Fin Formula

B.IV, D.1/7, P.1/11, V.6, A.32 (3/29), C.17

Comments

(*a*) Length of head 5.8-6.5; of caudal fin 4.5 and height of the body in 6-7 in the total length (15.4 to 23.8 cm).

(*b*) *Eyes*: With a broad circcular adipose lid situated nearly on the lower surface of the head, diameter 4-4.5 in the interorbital width.

(*c*) *Barbels*: 4 pairs, nasal barbels is very short and is ¼ length of head; the maxillary extending upto the middle of ventral fin, the outer mandibular barbel is about as long as the head while the inner pair is shorter.

(*d*) *Teeth*: Palate is with a samilunar band of teeth across. Vomerine and palatine patches of teeth on either side are continous.

(*e*) *Fins*: Dorsal fin arises is advance of ventral; the spine is slender and ruguse anteriorly while feebly serrated posteriorly.

(*f*) *Colour*: Greyish along the back and silvery along the side abdomen.

(*g*) *Distribution*: Fresh water of Northern part of India, Pakistan & Burma.

Glyptothorax madraspatanum

Classification

Series	:	**Pisces**
Class	:	**Teleostomi**
Subclass	:	**Actinopterygii**
Order	:	**Cypriniformes**
Division	:	**Siluri**
Suborder	:	**Siluroide**
Super family	:	**Siluroidiae**
Family	:	**Siluridae**
Genus	:	***Glyptothorax***
Species	:	***madraspatanum***

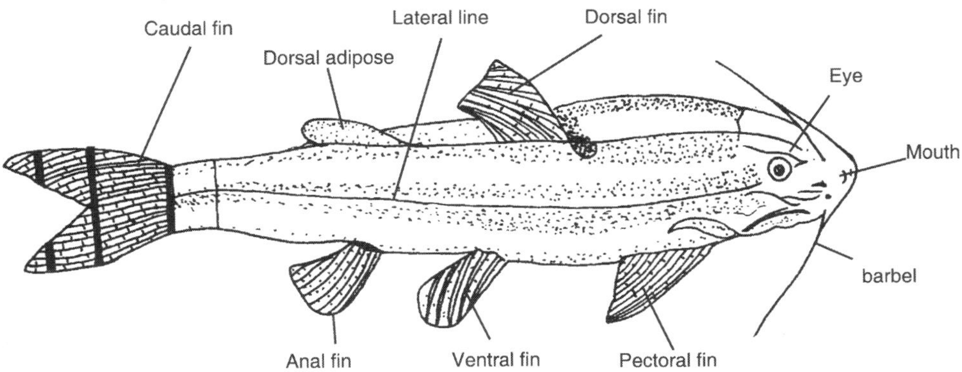

Fin Formula

B.IV, D.1/6/0, P.1/10, V.6, A.10-11(2/8-3/9), C.17

Comments

(*a*) Length of head 5-5¼; of caudal 4 1/3 - 4 2/3 height of body 6.5 in total length (6 to 7.0 cm).

(*b*) *Eyes*: Small, situated in the middle of the commencement of the posterior half of the head, the distance between the eyes equals 2/7 of the length of the head. The greatest width of the head rather rough, upper jaw the longer the width of the mouth equals 2/5 of the length of the head, lips not fringed. Occipital process very narrow, four times as long as its base.

(*c*) *Barbels*: The maxillary reach the base of the pectoral the nasal 2/3-3/4 of the distance to the orbit; the outer mandibular ones to the gill openings, whilst the internal are little shorter.

(*d*) *Teeth*: Generis

(*e*) *Fins*: Dorsal spine rather strong, not enveloped in skin, serrate posteriorly, and in some specimens anteriorly in its upper third, it is as long as the head behind the angle of the mouth, adipose rather longers than that of the rayed dorsal fin. Pectoral almost reaches the ventral, its spine not plaited, moderately strong and 3/4 as long as the head. Caudal peduncle-1/2 as high as long: skin smooth.

(*f*) *Colour*: Yellowish with dark bands; fin also yellow with black bands.

Glyptothorax carnirostris

Classification

Series	:	**Pisces**
Class	:	**Teleostomi**
Subclass	:	**Actinopterygii**
Order	:	**Cypriniformes**
Division	:	**Siluri**
Suborder	:	**Siluroidei**
Super family	:	**Silurodiae**
Family	:	**Siluridae**
Genus	:	***Glyptothorax***
Species	:	***carnirostris***

Fin Formula

B.IV, D.1/6/0, P.1/9, V.6, A.11-12 (2/9-2/10), C.17

Comments

(*a*) *Head*: Length of head 5 1/4; of caudal fin 5 1/2; height of body 5 1/2 in the total length.

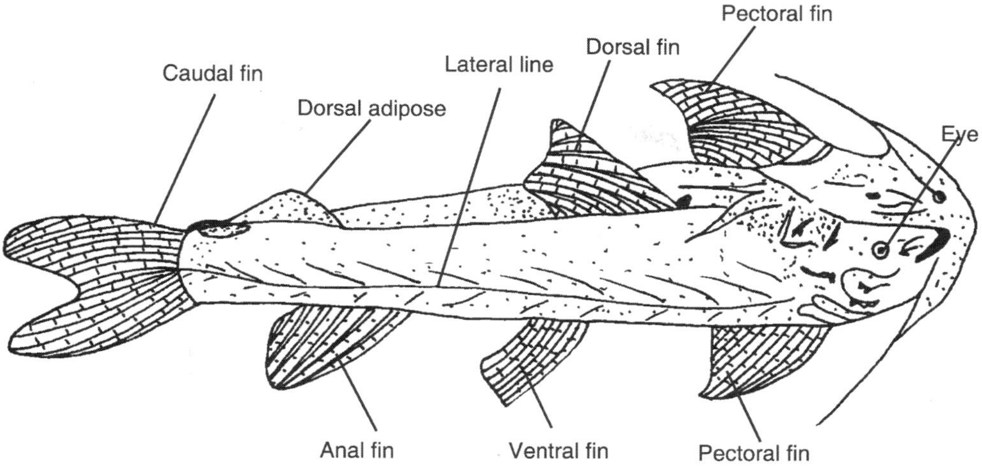

(*b*) *Eyes*: Slightly behind the middle of the length of the head, the width of the interorbital space equals 1/4 of the length of the head. The width of the head equals 2/3 of its length. Upper jaw the longer; the width of the mouth gap equals 1/2 of length of the head. Lips not the fringed. Occipital process three times as wide as long.

(*c*) *Barbels*: The maxillary have broad bases and extend to the middle of the pectoral fin. The nasal, which have a very wide frings, reach the hind edge of eye; the outer mandibular, which is also broad, to the gill opening, whilst the inner are shorter.

(*d*) *Teeth*: Generic.

(*e*) *Fins*: Dorsal as high or rather higher than the body; length of the base of the adipose dorsal exceeds that of the rayed fin. Pectoral reaches nearly to the ventral, its spine strong as long as that of the dorsal and leaving 13 denticulations internally, its not striated inferiorly. Ventral does not quite reach the anal, caudal forked.

Rita rita

Classification

Series	:	**Pisces**
Class	:	**Teleostomi**
Subclass	:	**Actinopterygii**
Order	:	**Cypriniformes**
Division	:	**Siluri**
Suborder	:	**Siluroidei**
Super family	:	**Silurodiae**
Family	:	**Bagridae**
Genus	:	***Rita***
Species	:	***rita***

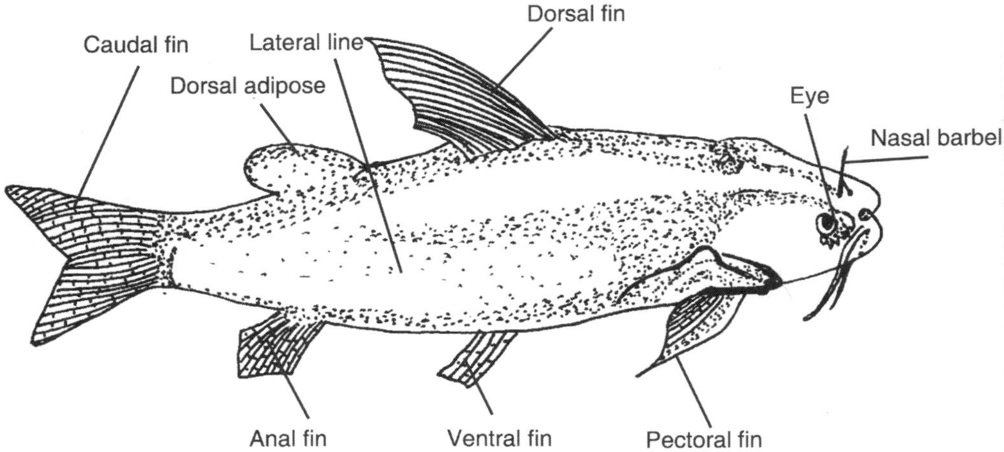

Fin Formula

D.16/0, P.1/10, V.8, A.13(4/9), C.19, B.III

Comments

(a) *Head*: Length of head 4.05 (4-4.1) and caudal fin 5 and height of body is 5.5 in the total length (5.6 to 7.6 cm).

(*b*) *Eyes*: Dia. 9.75 (9.5-1000) in the length of head the greatest width of head is equal to its length behind the 1st nostril occipital process, seapular, cubito humeral are granular. Creipital process as long as wide at its base appears longer but a part of base curved by skin. Biii nasal barbel short pointed tip, maxillary barbel near the head, mandibular little behind.

(*c*) *Teeth*: Villiform in upper jaw and out. Portion of lower jaw. Internally 2-3 rows of rounded teeth in lower jaw.

(*d*) *Fins*: Dorsal & Pectoral supported by the strong bony spines. Dorsal spine is lightly posteriorly on its upper part, length varies according to age, Base of adipose 6 of the rayed fin and the interspace between 2 fins pectoral spine shorter than Dorsal & denticulated inner margin. Ventral arise just behind the last dorsal ray. Caudal fin forked.

(*e*) *Colour*: Brown and lighter along the sides of abdomen.

(*f*) *Distribution*: Rivers of North part of India, Pakistan & Burma.

(*g*) *Local Name*: Hunna & Rita (Length 5.6-7.6 cms).

Mystus tengara

Classification

Series	:	**Pisces**
Class	:	**Teleostomi**
Subclass	:	**Actinopterygii**
Order	:	**Cypriniformes**

Division	:	**Siluri**
Suborder	:	**Siluroidae**
Family	:	**Bagridae**
Genus	:	***Mystus***
Species	:	***tengara***

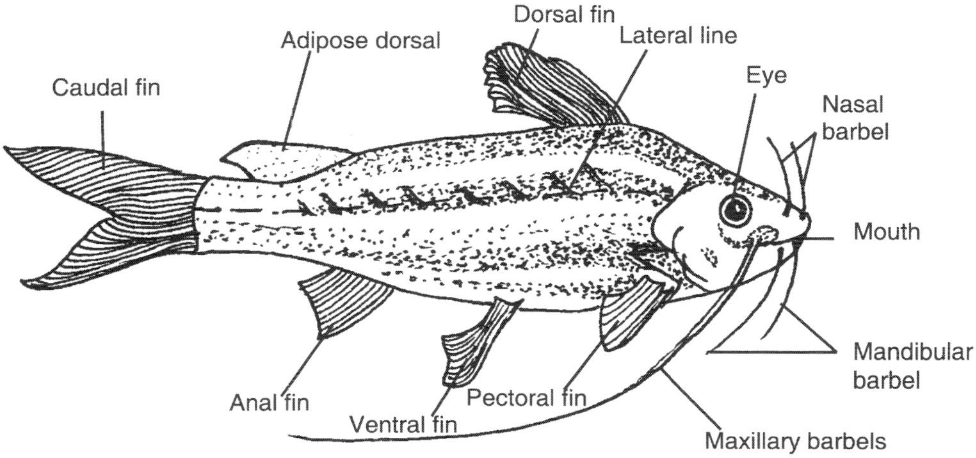

Fin Formula

D.1/7/0, P.1/8, V.6, A.12-13(2-3/9-10), C.19, Barbels 4 pairs

Comments

(*a*) *Head*: Length of head 4.5; of caudal fin 4.3 and the height of the body 4 in the total length (7.2 cm).

(*b*) *Eyes*: Dia. 3.95 (3.4-4) in the length of the head 1.3(1-1.5) from the end of snout and 1.57(1.4-1.75) in the interorbital width.

(*c*) *Barbels*: 4 pairs, nasal nearly as long as head, maxillary on reaches to the base of ventral fin. Ext. mandibular barbel reaches the middle of pectoral fin. Int. barbel comparatively shorter reaching up to base of pectoral.

(*d*) *Teeth*: A narrow band across palate.

(*f*) *Fins*: Dorsal spine slightly serrated anteriorly. It is as long as head excluding the snout and serrated posteriorly along its entire length.

(*g*) *Colour*: Brilliant yellow with black shoulder spots and 5 black longitudinal blands of which 4 on each side & 5th long the mid dorsal line hense common to both sides. Mandibular barbel with black streak base of ventral fin has rounded black blotch, while the base of and fin has an elongated faint black blotch.

(*h*) *Distribution*: Northern part of India & Pakistan.

(*i*) *Local Name*: Tengara.

Mystus cavasius

Classification

Series	:	**Pisces**
Class	:	**Teleostomi**
Subclass	:	**Actinopterygii**
Order	:	**Cypriniformes**
Division	:	**Siluri**
Suborder	:	**Siluroidae**
Family	:	**Bagridae**
Genus	:	***Mystus***
Species	:	***cavasius***

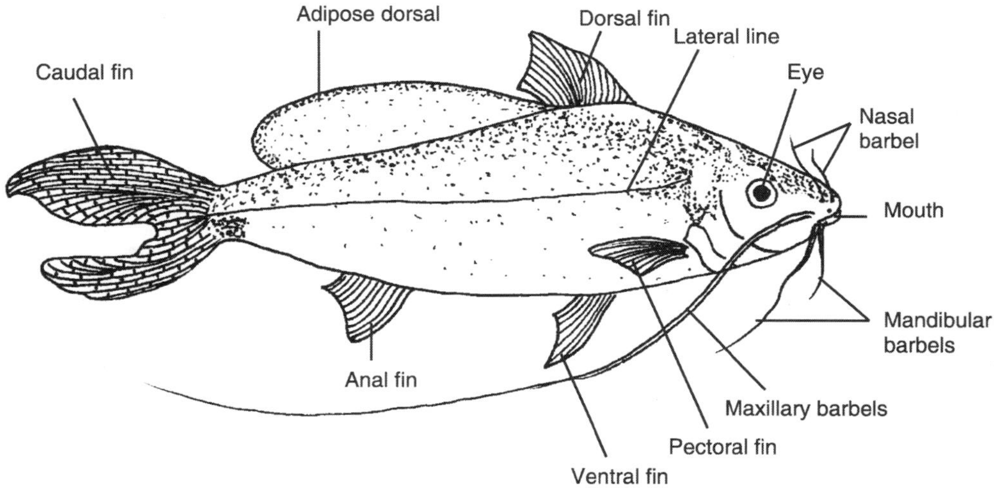

Fin Formula

B.VI, D.1/7/0, P.1/8, V.6, A.11-13 (4/7-9), C.16

Comments

(*a*) *Head*: Length of head 5.5-5.6; of caudal fin 6; height of body 5.23-6.5, & width of head is 7.5 to 8.5 in the total length (11.8 to 13.5 cm).

(*b*) *Eyes*: Dia. 3.5–4 in the length of head; 1.25-1.33 from the end of snout and 1 in the interorbital width.

(*c*) *Barbels*: 4 pairs, nasal barbels as long as the head; the maxillaries extend beyond the base of caudal fin; external mandibular barbles extend almost up to the base of ventrals which the internal mandibulars as long as the head.

(*d*) *Teeth*: Both the jaw carry teeth, those on the palatine are in one continuous crescentric band.

(*e*) *Fins*: Dorsal spine, higher than the body, is entire, pointed as long as the head excluding the snout. The adipose dorsal commence just behind the rayed one is fairly long, its base is nearly 3 times as long as the base of the rayed dorsal.

(*f*) *Distribution*: Northern part of India, Pakistan & Burma, in the south it extends up to Chilka lake and Madras.

Mystus seenghala

Classification

Series : **Pisces**
Class : **Teleostomi**
Subclass : **Actinopterygii**
Order : **Cypriniformes**
Division : **Siluri**
Suborder : **Siluroidae**
Family : **Bagridae**
Genus : ***Mystus***
Species : ***seeghala***

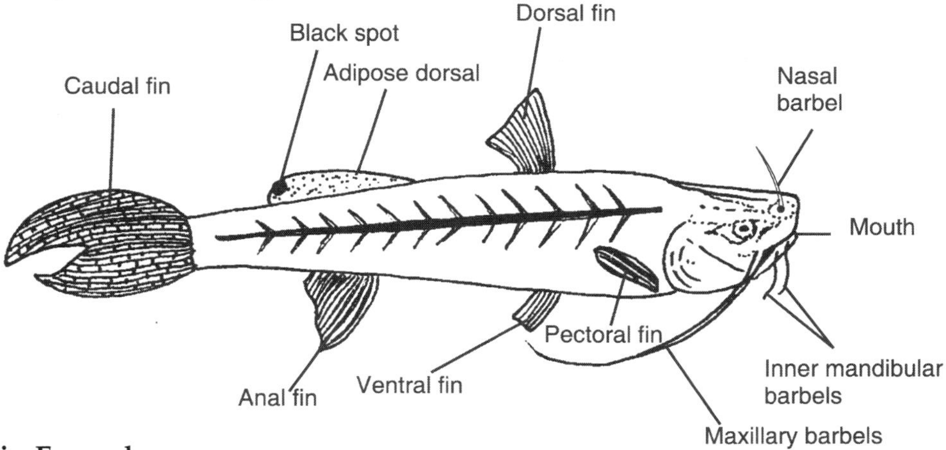

Fin Formula

B.XII, D.1/7/0, P.1/9, V.6, A.11-12(3/8-9), C.19-21

Comments

(*a*) *Head*: Length of head 4.5; of caudal fin 4.25, height of body 8 and the width of head is 8 in the total length (17.8 cm).

(*b*) *Eyes*: Dia 7.5 in the length of head, 2.5 from the end of snout and 1.5 in the interorbital width, snout is spatulate, upper jaw is linger, its width 1/3 of head length. Cleft of mouth is shallow, upper surface of head are graduated in ridge. The median longitudinal groove on the head reaches the base of occipital process.

(c) *Barbles*: 4 pairs maxillary extend just behind the hind margin of the dorsal fin or slighty beyond but never reach to the caudal fin. The nasal barbel reach opposite the middle of the orbit, the external mandibular reach to bone of pectoral fin while the internal mandibular 1/3 shorter.

(d) *Teeth*: On the palate in a continuous cresentric band.

(e) *Fins*: Dorsal spine weak, rugose anteriorly weakly serrated posteriorly is and long as the hind bone of the adipose, dorsal is equal to rayed fin. The intermediate is about the same length. Pectoral spine is stronger than the dorsal and is as long as the head, extremely denticulated with 20 teeth. Ventral fin with upper linger lobe deeply forked.

(f) *Distribution*: India, Pakistan and Burma. In India it extends to river Krishna only in South.

Nemacheilus savona

Classification

Series	:	**Pisces**
Class	:	**Teleostomi**
Subclass	:	**Actinopterygii**
Order	:	**Cyprinifor**
Division	:	**cyprini**
Family	:	**Cobitidae**
Subfamily	:	**Cobitini**
Genus	:	***Nemacheilus***
Species	:	***savona***

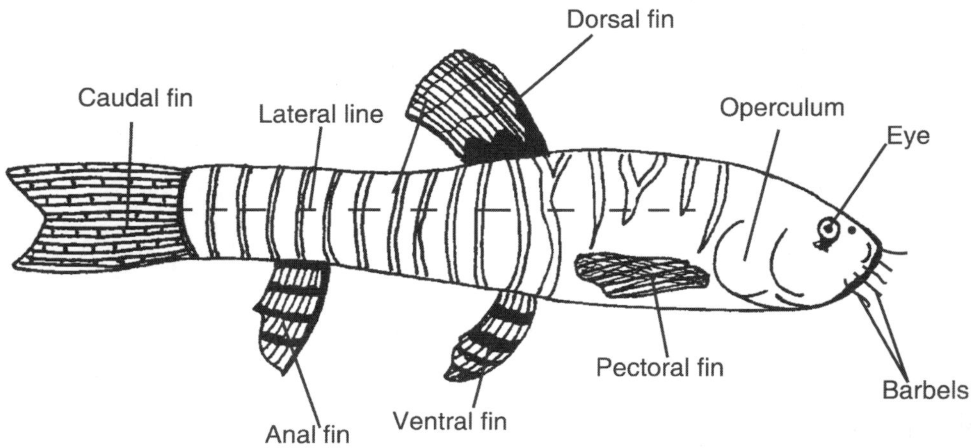

Fin Formula

B.III, D. 10-11 (2-3/8), P.10, V.7, A.7 (2/5), C.18

Comments

(*a*) *Head*: Length of head 4.5-4 3/4 of caudal 5.5–6 height of body 6 in the total length.

(*b*) *Eyes*: Rather large, rather behind the middle of length of head. Head as broad as it is long without snout. Barbel long, nostral maxillary one reach to the eye.

(*c*) *Fins*: Dorsal with its upper edge rather convex, it arises midway between the anterior edge of orbit and the base of caudal fin and its commencement before the origin of the ventrals. Pectoral extends 2/3 of the distance to ventral, which lasts reaches half way to the anal caudal (very slightly emerginate its lobe being rounded.

(*d*) *Lateral Line*: Complete.

(*e*) *Scales*: Small, most distant in the posterior portion of body.

(*f*) *Colour*: Purplish becoming lighter on the abdomen, having 10-12 very narrow vertical white bands not above 1/8 or 1/6 as wide as ground. A black band at the roof of caudal fin, a black blotch at the base of the few dorsal rays on to which the white body bands continued and 4-5 rows of dark spots on the fin, caudal with narrow bands of dark spots, 2 bands also is the ventral and anal fin.

(*g*) *Distribution*: Bengal & North-West provincess.

Nemacheilus montanus

Classification

Series	:	**Pisces**
Class	:	**Teleostomi**
Subclass	:	**Actinopterygii**
Order	:	**Cypriniformes**
Division	:	**Cyprini**
Family	:	**Cobitidae**
Sub family	:	**Cobitini**
Genus	:	***Nemacheilus***
Species	:	***montanus***

Fin Formula

B.III, D.9–10(8/7-8), P.10, V.7, A.7(2/5), C.19.

Comments

(*a*) *Head*: Length of head 3.5–3.6 of the caudal 3–4 height of the body 8.5 in total length.

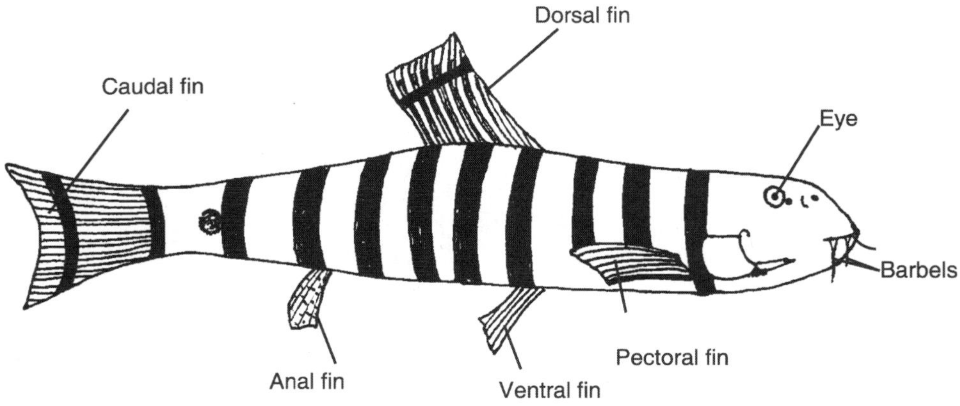

(*b*) *Eyes*: Small, dia 1/5 of the length of the head and situated before the middle of its length.

(*c*) *Barbels*: Short but snout.

(*d*) *Fins*: Dorsal commences midway between posterior margin of orbit and the base of the caudal fin. Peetoral reaches above half way of the root of the ventral, caudal with rounded lobes.

(*e*) *Scales*: Minute.

(*f*) *Colour*: 11 verticle bands of brown colour are present, on the body caudal with a black bands at its base and a bar across each lobe.

(*g*) *Habitat*: Himalayas, attaining the least 4.5" is length.

Nemacheilus rupicola

Classification

Series	:	**Pisces**
Class	:	**Teleostomi**
Subclass	:	**Actinopterygii**
Order	:	**Cypriniformes**
Division	:	**Cyprini**
Family	:	**Cobitidae**
Sub family	:	**Cobitini**
Genus	:	***Nemacheilus***
Species	:	***rupicola***

Fin Formula

B.III, D.9-11(2-3/7-8), P.10,V.8, A.7(2/5), C.18

Comments

(a) *Head*: Length of head 5.5; of caudal 6, height of body 7-8 in total length.

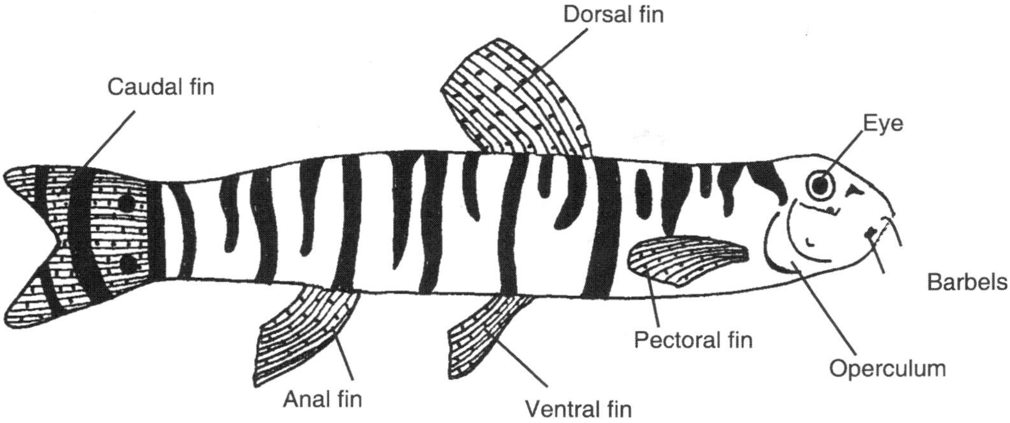

(*b*) *Eyes*: Small, in the middle of the length of head; the width of the interorbital space equal about 3/4 the length of the snout. Head nearly as wide as long. Mouth transverse and snout obtuse in the adult free portion of tail not as deep as long.

(*c*) *Fins*: The dorsal upper margin obligue it commences midway between the nostrils and the base of the caudal fin. Pectoral extends a little above half the distance to the ventral caudal emerginate.

(*d*) *Scales*: Absent or minute in the posterior half of the body.

(*e*) *Lateral line*: Complete.

(*f*) *Colour*: Yellowish, with 12-17 verticle brown bands much narrower than the intermediate ground colour, and decreasing in width below the lateral line, they are sometimes incomplete; a wide darkband across the base of caudal, fin as well as the dorsal has several rows of spots forming bands: a dark mark at the base of the anterior dorsal rays, sometimes another at the base of middle ones.

(*g*) *Distribution*: In Himalayas at 1600 feet above the sea level.

Nemacheilus botia

Classification

Series : **Pisces**
Class : **Teleostomi**
Subclass : **Actinopterygii**
Order : **Cypriniformes**
Division : **Cyprini**
Family : **Cobitidae**
Sub family : **Cobitini**
Genus : ***Nemacheilus***
Species : *botia*

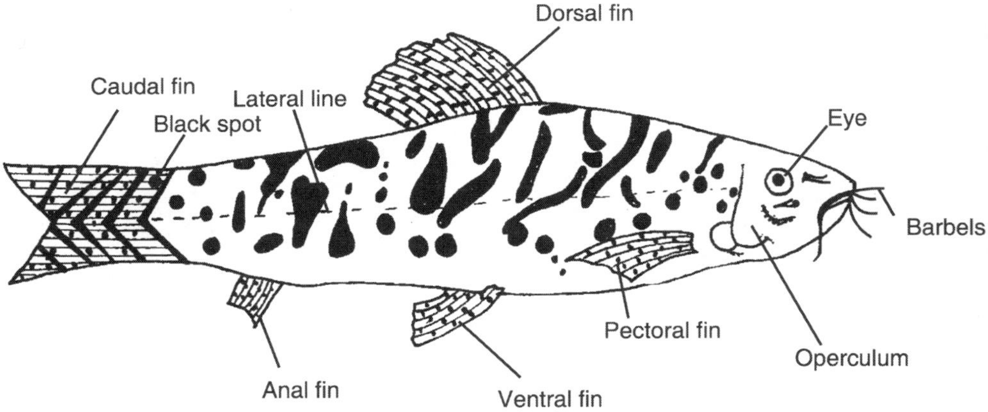

Fin Formula

D.12-14(2/10-12), P.11, V.8, A.7(2/5), C.17, Barbels 3 pairs

Comments

(*a*) *Head*: Length of head 2.5 to 5.5, of caudal 4.5, height of body 4¾ in the total length (8.0 cm).

(*b*) *Eyes*: Diameter 3.75-4 in the length of head, 1.5-1.75 Diameter from end of snout and 1.25 apart. The greatest width of the head equals half its length preorbital has in some specimens a projection which is movable below the orbit, whilst in others it is entirely concealed by the skin.

(*c*) *Barbels*: Long, the maxillary pair reaching to below the posterior edge of the eye.

(*d*) *Fins*: Dorsal commences rather nearer the snout than to the base of caudal fin, whilst the length of its base equals that of the head, its upper edge nearly straight. Pectoral as long as the head. Ventral inserted under middle of the Dorsal, caudal slightly notched.

(*e*) *Scales*: Distinct.

(*f*) *Lateral Line*: Complete, 12 rows of scales between it and the base of the ventral fin.

(*g*) *Colours*: Grayish 2 from 10-14 short bears on lateral line, and a number of irregular blotches above it, sometimes forming bands over the back. Dorsal fin orange and with rows of black spots; Caudal with about 7 irregular bars of a greater shape, and a black ossellus on the upper portion of the base of the fin.

(*h*) *Distribution*: Found throughout India and Ceylon except Malabar and south of river Krishna.

Puntius sophore

Classification

Series	:	**Pisces**
Class	:	**Teleostomi**
Subclass	:	**Actinopterygii**
Order	:	**Cypriniformes**
Division	:	**Cyprini**
Sub Order	:	**Cyprinoidei**
Family	:	**Cyprinoidae**
Sub family	:	**Cyprinimi**
Genus	:	***Puntius***
Species	:	***sophore***

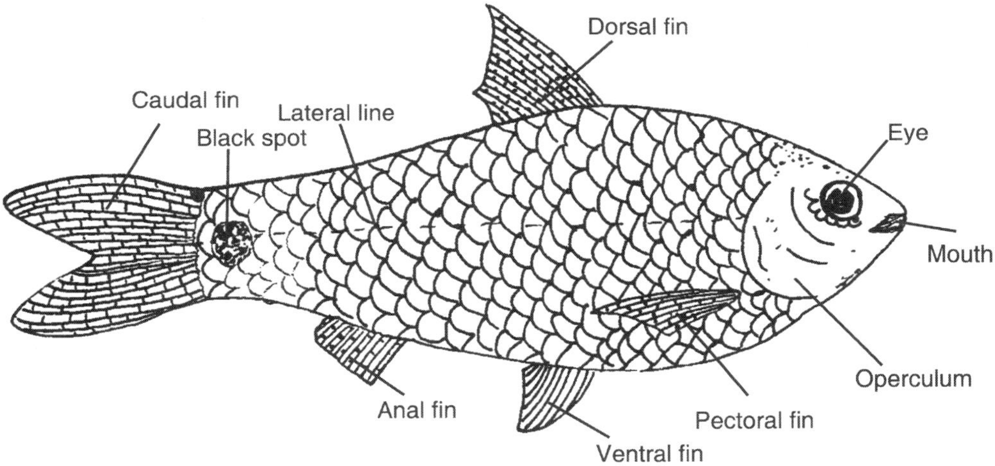

Fin Formula

D.11(3/8), P.15-16, V.9, A.8(3/5), C.19, L.I. 25-26, L.t.r. 5–5.5/5–5.5.

Comments

(*a*) Length of head 5, of caudal 5, and height of body 3.62 (3.5-3.75) in the total length.

(*b*) *Eyes*: Dia. 3.7 (3.5-4) in the length of head, 1 from the end of snout and 1.25 (1.3-1.4) in the interorbital width.

(*c*) *Barbels*: Absent.

(*d*) *Fins*: Dorsal inserted opposite the insertion of ventral midway between the snout and the base of caudal fin and 1/2-2/3 as high as body. Its last undivided ray is moderate strength, osscous entire as long as head without snout or lers, caudal fin forked.

(*e*) *Lateral Line*: Complete 3.5 row of scales, between it and the base of the ventral fin, a row of 9 predorsal scales.

(*f*) *Colour*: Silvery and the dorsum darker. Only in breeding season appears a scarlet red lateral band extending from the base of gill to the base of caudal fin. Black band placed along the lateral line. In preserved specimen this band fades and finally disappears. There are two black blotches one at the base of 3-5th branched dorsal ray & outer at the base of caudal fin over 22nd & 23rd scales along the lateral line. In gavemile caudal blotch found absent.

(*g*) *Distribution*: Fresh & brackish waters of India, Pakistan and Burma.

(*g*) Total length of the fish 6.3 to 10.4 cm.

Chagunius chagunio

Classification

Series	:	**Pisces**
Class	:	**Teleostomi**
Subclass	:	**Actinopterygii**
Order	:	**Cypriniformes**
Division	:	**Cyprini**
Suborder	:	**Cyprinoidci**
Family	:	**Cyprinidae**
Sub family	:	**Cyprinini**
Genus	:	***Chagunius***
Species	:	***chagunio***

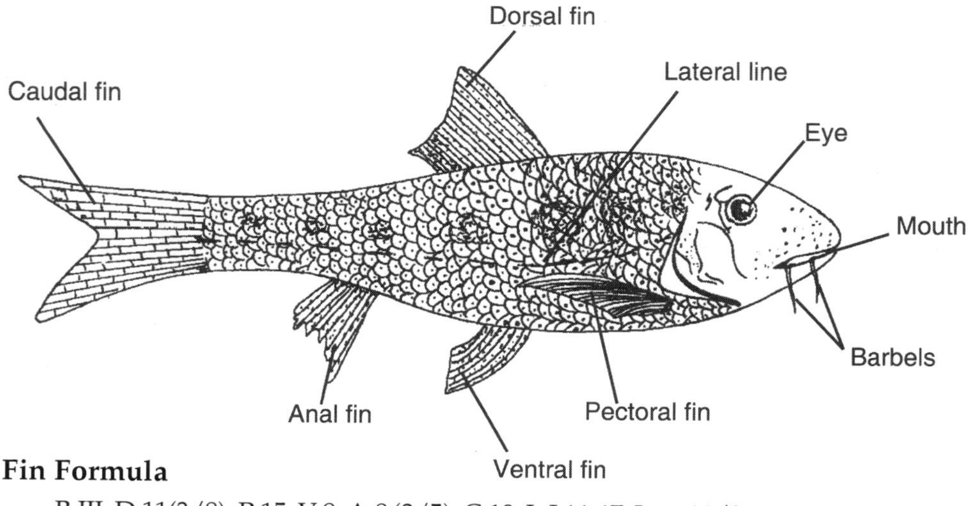

Fin Formula

B.III, D.11(3/8), P.15, V.9, A.8 (3/5), C.19, L.I 44-47, L.tr. 11/9

Comments

(*a*) Length of head 4.75; of caudal 5 and height of body 4 in the total length (24.2 cm).

(*b*) *Eyes*: Diameter 5 in the length of head, 2.3 from the end of snout and 1.8 in the interorbital width.

(*c*) *Barbels*: A pair of maxillary and a pair of rostrals, both pair longer than orbit.

(*d*) *Fins*: Dorsal commences midway between the end of snout and the base of caudal fin, its last undivided ray is osseous, strong and width coarse teeth. Caudal forked.

(*e*) *Lateral line*: Complete 6 rows of scales between it and the base of the ventral fin. A row of 17 predorsal scales.

(*f*) *Colour*: Uniform silvery with a pinkish tinge. The summit of dorsal fin, is black edges of scales with black dots.

(*g*) *Distribution*: Northern India and Pakistan.

Crosscheilus latius

Classification

Series	:	**Pisces**
Class	:	**Teleostomii**
Sublcass	:	**Actinopterygi**
Order	:	**Cypriniformes**
Division	:	**Cyprini**
Suborder	:	**Cyprinoidei**
Family	:	**Cyprinidae**
Subfamily	:	**Cyprinini**
Genus	:	***Crossocheilus***
Species	:	***latius***

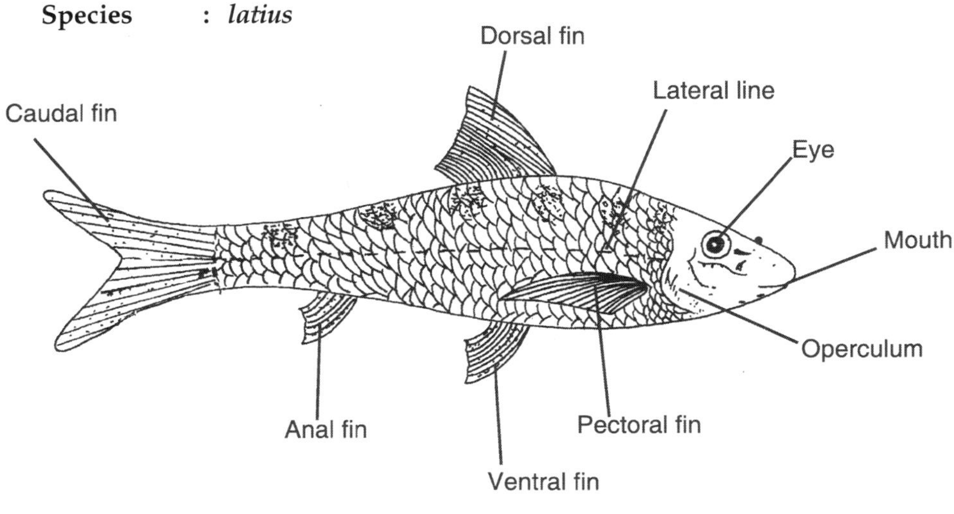

Fin Formula

D.11(3/8), P.15, V.9, A.7(2/5), C.19, L.I.39.

Comments

(*a*) *Length*: Length of head 6.33 (6-6.5); of caudal 5, height of body 6 and the height of the caudal peduncle is 12.55 (12 to 12.5) in the total length.

(*b*) *Eyes*: In the middle of the head; Dia 3.2(3 to 3.5) in the length of head' from the end of snout and 1 is the interorbital width upper surface of the head is broad, snout over hanging the jaws and is devoid of any lobe. Upper lip is indented on the edge. Lower lip carries a horny covering.

(*c*) *Barbles*: 2 pairs, a rostral pair and a shorter maxillary pair.

(*d*) *Teeth*: Pharyngeal are crooked 5,4, 2/2, 4, 5.

(*e*) *Fins*: Dorsal fin commences nearer the smount than the base of caudal fin. pectoral fin is as long as the head. And the caudal fin is deeply forked with its upper lobe a bit longer than the lower.

(*f*) *Scales*: 4.5 rows between lateral line and the base of the ventral fin.

(*g*) *Colour*: White with slight greenish tinge along the upper 1/3 which is irregularly marked with black spots; getting silvery white present over the occipnt.

(*h*) *Distribution*: Fresh waters of Northern part of India all along Himalayas.

(*h*) The total length of fish–4.4 to 6.5 cm

Remark: It is a torrential fish as it is seen in this locality only during monsoon.

Danio devario

Classification

Series	:	**Pisces**
Class	:	**Teleostomi**
Subclass	:	**Actinopterygii**
Order	:	**Cypriniformes**
Division	:	**Cyprini**
Suborder	:	**Cyprinoidei**
Family	:	**Cyprinidae**
Subfamily	:	**Cyprinini**
Genus	:	***Danio***
Species	:	***devario***

Fin Formula

D.18 (2/16), P.14, V.8, A.19 (3/16), C.19, L.I.49, Ltr. 11.5/2.5

Comments

(*a*) *Head*: Length of the head 5 to 5.2, caudal 4.5 and height of body 3 to 3.2 in the total length.

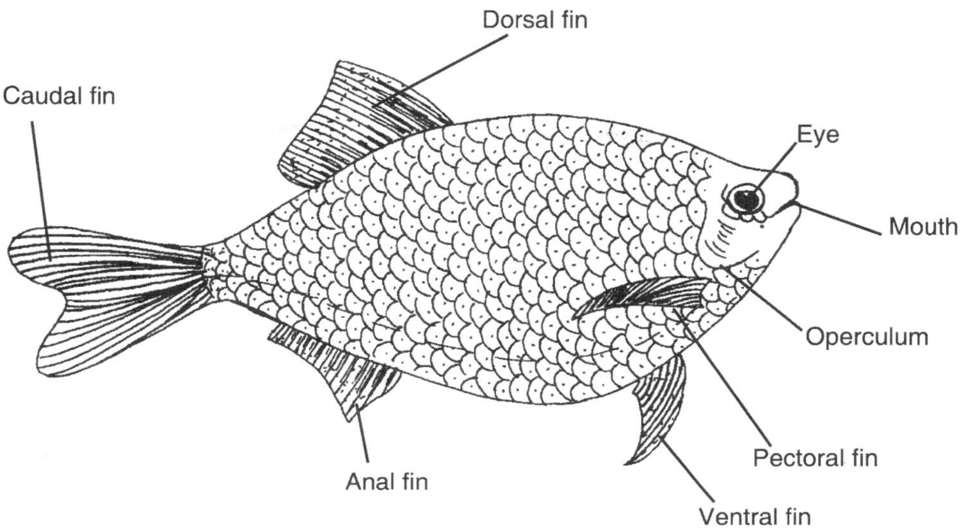

(b) *Eyes*: Diameter 3 in the length of the head; 0.8 from the end of snout and 1.52 in the interorbital width.

(c) *Barbles*: Absent.

(d) *Fins*: Dorsal fin commences midway between the anterior margin of the orbit and the base of the caudal fin. Caudal fin lunate.

(e) *Scales*: 2.5 rows between lateral line and the base of the ventral fin. These are 16 predorsal scales.

(f) *Colour*: Greenish above and dull white below. A bluish band along the mid lateral line becomes broad and prominent posteriorly. There is a black spot at the posterior margin of head.

(g) *Distribution*: Fresh waters of N. India & Pakistan.

(g) Total length–7.3 to 7.7 cm.

Barilius vagra

Classification

Series	:	**Pisces**
Class	:	**Telepstomi**
Subclass	:	**Actinopterygii**
Order	:	**Cypriniformes**
Division	:	**Cyprini**
Suborder	:	**Cyprinoides**
Family	:	**Cyprinidae**
Sub family	:	**Cyprinini**
Genus	:	***Barilius***
Species	:	***vagra***

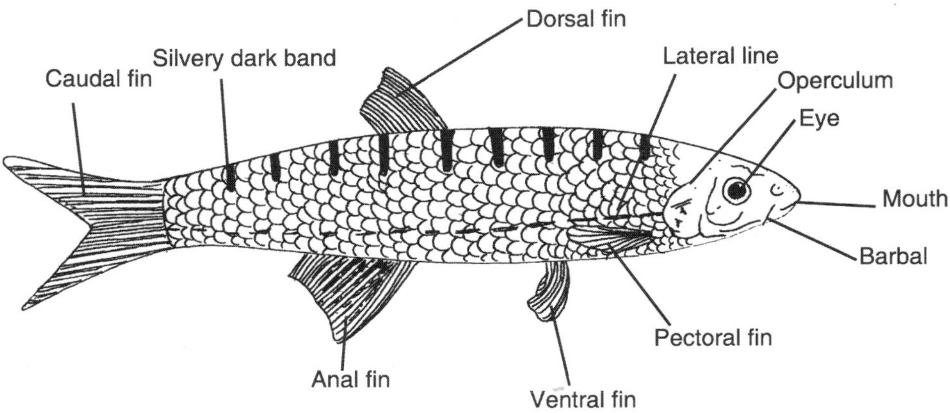

Fin Formula

B.III, D. 9(2/7), P.16, V.9, A.13-15 (2-3/11-12), C.19, L.I. 42-44, L.tr. 7-8/4, Vert. 16-28.

Comments

(*a*) *Head*: Length of head 5-5.5 of caudal 5 height of body 5-6 in the total length (15 cm).

(*b*) *Eye*: Dia. 3.5-4 in the length of the head, 1 dia., from the end of snout & 1.25 apart. The post extremity of maxilla extends to beneath the shape of the middle of the orbit. Lower jaw strongly longer, 3rd sub orbital body bone as deep as the uncovered portion of the check below it jaws of equal length anteriorly, upper sometimes a little notched. The lower curved with large pores in adult. Axillary process very short.

(*c*) *Barbles*: Nostral nearly half long as head. Maxillary pair very short.

(*d*) *Teeth*: Pharyngeal crooked 5, 3, 2/2, 3, 5.

(*e*) *Fins*: Dorsal higher than long commencing midway between the post edge of the orbit and the roof of the caudal. Its last 2 rays being over the anal, pectoral not so quite long a head. Caudal deeply forked.

(*f*) *Lateral line*: 2-2.5 rows of scales between it and base of ventral fin 20 rows before the base of dorsal fin.

(*g*) *Colours*: Silvery 10-14 dark bands sometimes indistinct, descends from the back of lateral line. Fins yellowish, edge of caudal stained grayish.

(*h*) *Distribution*: Sindh hills, river of Himalayas and sub Himalayas range in Yamuna and Ganga also.

Chapter 7

Microbiological Analysis

(I) Introduction

Micro-organisms in water may be of two kinds:

1. Pathogenic
2. Non-pathogenic.

Non-pathogenic micro-organisms are harmless to human purposes and sometimes they may be beneficial for human consumption, where as pathogenic bacteria may cause several types of diseases such as cholera, typhoid, tuberculosis etc. So the presence of pathogenic or non-pathogenic bacteria in water decides the quality of water and its status for consumption for human beings whether it is for agricultural, industrial or drinking purpose.

Therefore we can say that the microbiological examination of any kind of water is necessary to know the status of its pollution before use. Some bacteria also acts as an indicator bacteria *i.e.* their presence confirms the presence of other type of bacteria. One of them is Coliform bacteria which are discharged from human intestine, their presence indicates the possibility of pathogenic bacteria.

In this chapter, we may describe the sampling process, prevention and storage of sample, preparation of various types of media for colony development, SPC/HPC estimation, estimation of MPN and estimation of aquatic fungi.

(*a*) Sampling Procedure

Extreme care should be taken in sample collection for microbial examination to avoid any type of contamination from other sources such as air. For this sterile bottles should be used. Before collection of sample bottles should be sterilized at 100°C in an oven for about 4-5 hours and should be wrapped in butter paper to avoid contamination. The bottles should be open at site and should be closed immediately after collection of samples. These samples should be opened in sterile conditions (*i.e.*- in laminar flow) and analyzed. While collecting the sample for microbial purpose, use sterile gloves to avoid contamination from hand.

(b) Preservation and Storage of Sample

It is very important that sample should be analyzed immediately after collection, if not possible keep the sample in ice to avoid any kind of change in microbiological parameters.

(c) Preparation of Growth Media

Growth Medium 1 (Nutrient Agar Medium NAM)

Beef Extract	1.0g
Yeast Extract	2.0g
Peptone	5.0g
NaCl	5.0g
Agar	15.0g
Distill Water	1.0 liter

Growth Medium 2 (Potato Dextrose Agar PDA)

Potato (Scrubbed and diced)	200g
Dextrose	20g
Agar	15g
Distill Water	1.0 liter
pH	5.6

Rinse potatoes under running water, add water and boil for about one hour, pass it through a fine sieve, squeeze through as much pulp as possible. Add Agar and boil until dissolved, remove from heat and add dextrose and stir to dissolve. Autoclave at 121°C/15 minute.

Growth Medium 3 (CM Medium)

Yeast extract	4.0g
Malt extract	10.0g
Glucose	4.0g
Agar	20.0g
Distill Water	1.0 liter
pH	7.2-7.4

Growth Medium 4 (Neo-peptone-glucose-rose-bengal aureomycin agar)

Neo-peptone	5.0g
Glucose	10.0g

Rose-bengal solution	3.5 ml
Agar	20.0gm
Distill Water	1 liter
pH	6.5

For preparation of Rose-bengal solution add 1gm rose bengal dye in 100ml of distilled water. Dissolve 1g chlortetracycline or tetracycline in 150 ml distilled water and refrigerate. Now add 0.05 ml of this solution to 10 ml rose-bengal aureomycin agar medium.

Growth Medium 5 (Czapeck agar)
For *Aspergillus, Penicillium, Paccilomyces*

Sucrose	30.0g
$NaNO_3$	3.0g
K_2HPO_4	1.0g
MgSO4. $7H_2O$	0.5g
KCl	0.5g
$FeSO_4$	0.01g
Agar	15.0 g
Distill water	1 liter

Growth Medium 6 (Sabouraud Agar)

Glucose	40.0g
Peptone	10.0g
Agar	20.0g
Distill Water	1.0 liter
pH	5.53

Growth Medium 7 (Malt Yeast Agar)

Yeast extract	3.0g
Malt extract	3.0g
Peptone	5.0g
Glucose	10.0g
Agar	20.0g
Distill Water	1.0 liter
pH	7.0

Growth Medium 8 (Yeast Glucose Agar)

Yeast extract	5gm
Glucose	10gm
Agar	15gm
Distill water	1.0 liter
pH	5.8

Growth Medium 9 (Neo-peptone-glucose-agar)

Neo-peptone	5.0g
Glucose	10.0g
Agar	20.0g
Distill water	1.0 liter
pH	6.5

Growth Medium 10 (Mac. Conkey broth)

Peptone	20.0g
Lactose	10.0g
Sodium Chloride	5.0g
Bile salts	5.0g
*Neutral red solution	1.0ml
Distill water	1.0 liter

***Neutral red solution**

Methyl red	0.1g
Ethyl alcohol	300ml
Distill water	200ml

Growth Medium 11 (Yeast Mannitol Salt Agar)

Mannitol	10.0g
Yeast extract	0.25g
$MgSO_4 . 7H_2O$	0.25g
$CaCl_2 . 2H_2O$	30mg
$Na_2HPO_4 . 12H_2O$	1.2g
KH_2PO_4	0.55g
NaCl	0.25g
$FeSO_4 . 7H_2O$	3.5mg
$ZnSO_4 . 7H_2O$	160mg
$CuSO_4 . 5H_2O$	80mcg

H_3BO_3	500mcg
$MnSO_4.4H_2O$	400mcg
Agar	18g
Distill water	1 liter
pH	7.0

Growth Medium 12 (Tryptone-glucose Yeast Agar)

Tryptone	5.0g
Yeast extract	2.5g
Glucose	1.0g
Agar	15.0g
Distill Water	1 liter
pH	7.0-7.2

Growth Medium 13 (RZA Agar)

Yeast extract	0.5g
Polypeptone	0.5g
Casamino acids	0.5g
Glucose	0.5g
Starch	0.5g
K_2HPO_4	0.3g
$MgSO_4.7H_2O$	0.05g
Sodium Pyruvate	0.3g
Agar	15.0g
Distill water	1 liter
pH	7.2

Growth Medium 14 (Potato Carrot Agar)

Potato	20g
Carrot	20g
Agar	20g
Distill water	1 liter

Growth Medium 15 (Malt extract Agar Blakerlee's formula)

Malt extract	20g
Glucose	20g
Peptone	1.0g
Agar	20g
Distill water	1.0 liter

Growth Medium 16 (BGLB)

Peptone	10g
Lactose	10g
Oxgall/Bile salt	20g
Brilliant green	0.0133g
Distill water	1 liter
pH	7.2±0.2

(d) Heterophic Plate Count (HPC)

HPC which was formerly known as Standard plate count is used to provide the density of bacteria in water sample.

Apparatus

1. Incubator
2. Petridishes
3. Conical flask
4. Biological pipettes
5. Colony counter
6. Water bath
7. Culture Tubes.

Two methods have been commonly used to get bacterial colonies these are:

(*a*) Pour Plate Method

In this method, six culture tubes each containing 9 ml distill water were sterilized after plugging for serial dilution. The sample was vigorously shaken and 1ml was transferred to the first culture tube. Now 1ml water from first tube was again transferred to second test tube and so on. Separate pipettes were used for each transfer. The dilutions were 1:10 or 10^{-1}, 1:100 or 10^{-2}, 1;1000 or 10^{-3}, 1:10000 or 10^{-4} and so on.

Now take 3 pairs of sterilized petridishes gently open the plate and transfer 1 ml of liquid from each dilution, add about 20ml of medium (NAM) into these plates. Gently rotate the plates to mix the medium and water sample. After solidification of medium invert the plate and incubate at 37.0 ± 2°C temperature for 48 hours.

(*b*) Streak Method

From these paired plates, individual colonies were transferred with the help of sterilized loop (by incineration), on media plates and streaked. These plates were incubated for further 48 hours at 37.0 ± 2°C.

(e) Most Probable Number (MPN)

The bacteriological quality of water is based on testing for bacterial species of known excretal origin, particularly organism of the coliform group. The coliform

Figure 7.1: Plates Showing Pouring and Streaking Techniques

group of bacteria includes all the aerobic and facultatively anaerobic, gram negative, non-speculating bacilli that produce acid and gas from the fermentation of lactose.

The multiple tube fementation technique can be used to enumerate positive presumptive and confirmed coliform test.

Apparatus & Material

1. Fermentation tubes (50ml)
2. Durham's vials
3. Biological pipettes
4. Incubator
5. Inoculation loop
6. Autoclave
7. Conical flask
8. Spirit lamp
9. Water bath
10. Laminar flow

A. Presumptive Test

Procedure

Prepare dilution above 1:1000, prepare 5 fermentation tubes for each dilution, previously filled with 9ml of Mac Conkey broth. Now put one Durham's vial inside each fermentation tube in an inverted condition. Plugged the tube with cotton and sterilize it in an autoclave. After autoclaving add sample using separate sterilized pipette for each test tube. Shake the tubes vigorously and incubate it at 35-37°C for 48 hrs. After 48 hrs. examine each tube carefully for gas production. The tubes showing gas in the Durham's vial are recorded as Positive(+Ve) even a tiny bubble should be recorded as +Ve.

The tube showing positive results were now subjected to the confirmatory test. Negative tubes were incubated for further 24 hrs. to confirm the test.

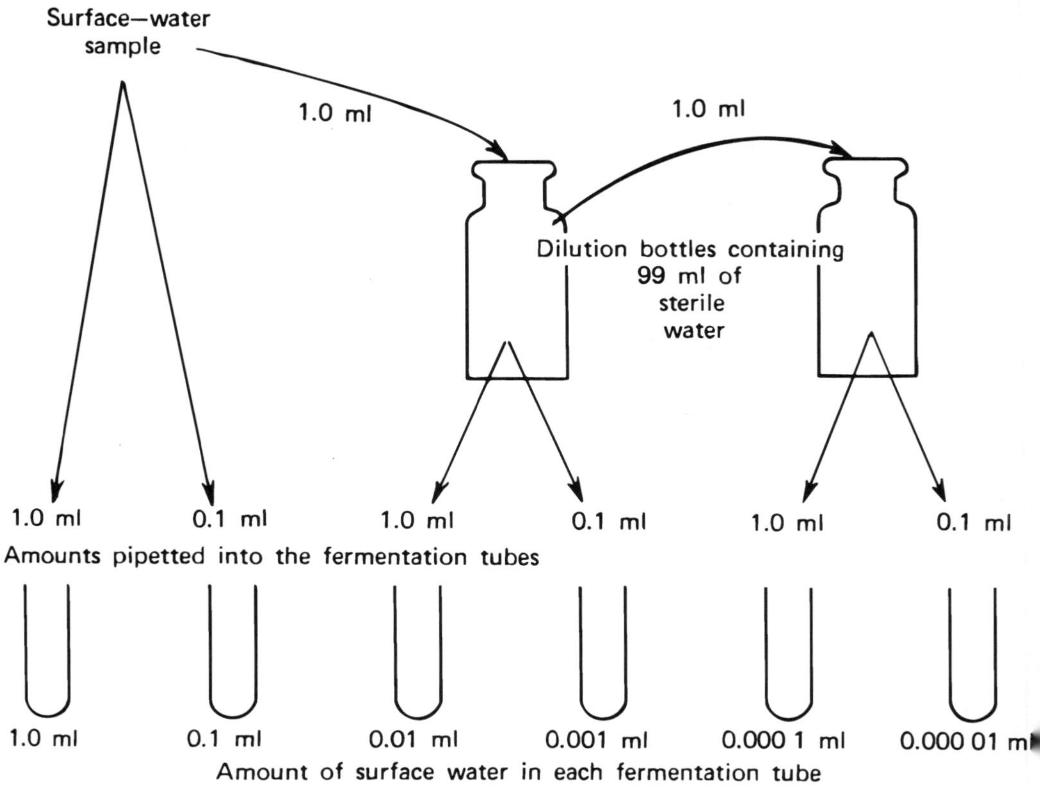

Figure 7.2: Preparation of Serial Dilution

B. Confirmatory Test

For confirmatory test, Brilliant green lactose bile (BGLB) broth is required. This inhibits the growth of lactose ferments other than coliform thus gas formation in BGLB broth constitutes a confirmed test *i.e.* coliform present.

Procedure

Fill the fermentative tube with 10ml BGLB broth, now put Durham's vial in an inverted position in each fermentation tube. The number of tubes should be equal to all positive tube in the presumptive test. Shook gently the positive tubes. Now transfer one loopful of sample to each fermentation tube having BGLB broth using inoculation loop. Now incubate these at 35-37°C for 48 hrs.

Examine tube after 48 hrs, if there is gas production in tube it is recorded as positive. Now put these positive number of tubes in MPN table given below and then in formula to calculate the MPN/100ml.

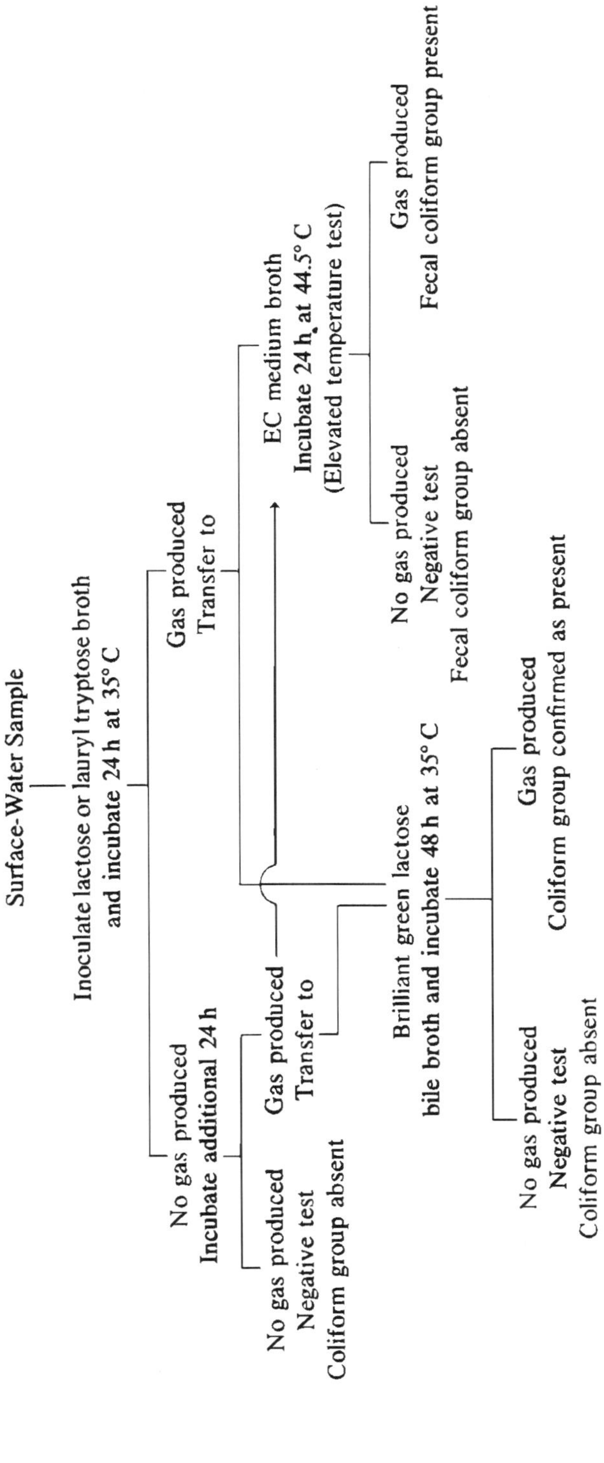

Flow Diagram 7.1: MPN Measurement

Table 7.1: MPN/100 ml for Various Combinations of Positive Results when 5 Tubes Each of 10, 1 and 0.1 ml Sample Fraction are Used

No. of Tubes Giving Positive Reaction out of			MPN Index per 100 ml	95% Confidence Limits		No. of Tubes Giving Positive Reaction Out of			MPN Index per 100 ml	95% Confidence Limits	
5 of 10 ml Each	5 of 1 ml Each	5 of 0.1 ml Each		Lower	Upper	5 of 10 ml Each	5 of 1 ml Each	5 of 0.1 ml Each		Lower	Upper
0	0	0	<2			4	2	1	26	9	78
0	0	1	2	<0.5	7	4	3	0	27	9	80
0	1	0	2	<0.5	7	4	3	1	33	11	93
0	2	0	4	<0.5	11	4	4	0	34	12	93
1	0	0	2	<0.5	7	5	0	0	23	7	70
1	0	1	4	<0.5	11	5	0	1	31	11	89
1	1	0	4	<0.5	11	5	0	2	43	15	110
1	1	1	6	<0.5	15	5	1	0	33	11	93
1	2	0	6	<0.5	15	5	1	1	46	16	120
						5	1	2	63	21	150
2	0	0	5	<0.5	13						
2	0	1	7	1	17	5	2	0	49	17	130
2	1	0	7	1	17	5	2	1	70	23	170
2	1	1	9	2	21	5	2	2	94	28	220
2	2	0	9	2	21	5	3	0	79	25	190
2	3	0	12	3	28	5	3	1	110	31	250
						5	3	2	140	37	340
3	0	0	8	1	19						
3	0	1	11	2	25	5	3	3	180	44	500
3	1	0	11	2	25	5	4	0	130	35	300
3	1	1	14	4	34	5	4	1	170	43	490
3	2	0	14	4	34	5	4	2	220	57	700
3	2	1	17	5	46	5	4	3	280	90	850
3	3	0	17	5	46	5	4	4	350	120	1000
4	0	0	13	3	31	5	5	0	240	68	750
4	0	1	17	5	46	5	5	1	350	120	1000
4	1	0	17	5	46	5	5	2	540	180	1400
4	1	1	21	7	63	5	5	3	920	300	3200
4	1	2	26	9	78	5	5	4	1600	640	5800
4	2	0	22	7	67	5	5	5	≥2400		

Source. Standard Methods for the Examination of Water and Wastewater.

Calculation

$$MPN/100ml = \frac{MPN\ table\ value \times 10}{Starting\ value}$$

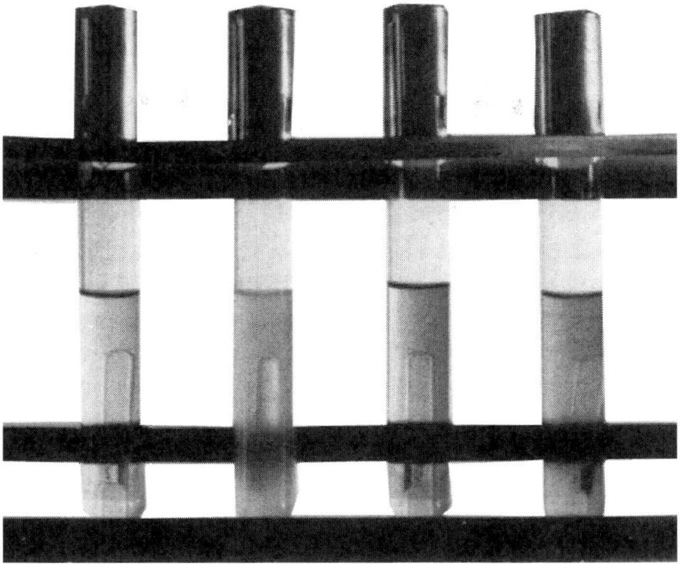

Figure 7.3: Tubes Showing Positive & Negative MPN Tests

(f) Aquatic Fungi

To find the number of aquatic fungi in a water sample, the procedure is same as described for HPC. In finding fungi Neo-peptone rosebengal aureomycin agar medium is used instead of NAM.

Chapter 8
Data and its Statistical Method/ Biostatistics

(a) Introduction

The application of statistical methods to the problems of biology is simply termed as biostatistics. According to Webster Statistics is defined as "The classified fact representing the conditions of the people in a state especially those facts which can be stated in number or in a table of numbers or in any tabular or classified arrangement".

According to Secrist Statistics is defined as "The aggregate of facts affected to a marked extent by multiplicity of causes, numerically expressed, enumerated or estimated according to reasonable standard of accuracy, collected in a systematic manner for a pre-determined purpose & Placed in relation to each other. "In other words statistics is defined as "The scientific study of numerical data based on natural phenomenon".

(b) Characteristics of Statistics

Statistics are the aggregate of facts & are numerically expressed. It should be collected in a systematic manner for pre-determined purpose. Statistics should be placed in relation to each other. And the reasonable standard of accuracy should be maintained in statistics. All these characteristics of statistics are according to Secrist.

(c) Uses of Statistics

(1) Large quantity of data should be presented in simpler form with the help of statistics.

(2) The relationship between variable is found with the help of statistics.

(3) Statistics provides various methods to compose data.

(4) Statistics is useful in various types of fields.

(d) Central Tendency and its Measures

The main aim to measure central tendency is to obtain one single value that describes the characteristics of whole set of data *i.e.* the single value which we obtain is the most representative of all the values. eg. In a class section A obtained an average marks of 70 per cent for boys and an average marks of 65 per cent for girls. Obviously not all the boys obtained 70 per cent marks and not all the girls obtained 65 per cent marks this average percentage provided the aggreagate percentage of boys and girls in a class. It is clear from the above example that an average represents the whole set of data or values therefore its values lies somewhere between the highest and lowest percentage of the whole class.

Definition

"That value of distribution which is considered as the most representative or typical value for a group".

Or

"Central tendency is a statistical measure that identifies the single most representative score for an entire distribution".

Or

According to Tate

"It is a sort of average or typical value of the items in the series and its function is to summarize the series in terms of this average value".

Central tendency can be measured by three ways:

(1) Mean

(2) Median

(3) Mode

All the above three measures are representatives of central tendency but have different characteristics. All the three representatives describes a single value for a whole set of data or values but one should have to decide which of the three measures is best for any particular set of data. Before analyzing the central tendency one should keep in mind that the main aim of central tendency is to obtain the single most representative value.

(i) Mean

The mean or arithmetic mean, is obtained by adding up all the values and dividing it by the total numbers of values.

"The mean for a distribution is the sum of scores divided by the number of scores".

Mean may be calculted by following ways:

(*i*) For Individual Observation

Add all the values and divide it by the number of observation. The mean is given by:

$$\bar{x} = \frac{\Sigma x}{n}$$

where,

$\bar{x}$: Mean

Σx: sum of all the variables

n: Number of variables

Example: Find the arithmetic mean of the dissolve oxygen observed in months of the year 2003.

7,8,6,9,10,7,6,8,9,8,6,10

Solution

$$\bar{x} = \frac{\Sigma x}{n}$$

$$\Sigma x = x_1 + x_2 + x_3 + - - - - x_{12}$$

$$\bar{x} = \frac{7+8+6+9+10+7+6+8+9+8+6+10}{12} = 7.83$$

$$\bar{x} = 7.83$$

(*ii*) When Frequencies are Given

Let x be the variables and f be the frequencies of any given set of data. Calculate fx by multiplying the variable and frequency now add values of fx to obtain Σf to calculate mean.

$$\bar{x} = \frac{\Sigma fx}{\Sigma x}$$

$$\bar{x} : 7 + 8 + 6 + 9 + 10 + 7 + 6 + 8 + 9 + 8 + 6 + 10$$

where,

$\bar{x}$: Mean

Σfx: Sum of all the values obtained after multiplication of variables and frequency

Σx: Sum of frequency.

e.g. From the given table calculate mean

Weight (x) gms	No. of Fish (f)	fx
50	10	500
70	8	560
80	12	960
60	15	900
	Σf = 45	Σfx = 2920

$$\bar{x} = \frac{\sum fx}{\sum f} = \frac{2920}{45}$$

$$\bar{x} = 64.88$$

(iii) Where Class Intervals are Given

Calculate the mid values of class intervals and multiply it with frequency to obtain fx, add values of fx to obtain Σfx and divided it by Σf to calculate mean

$$\bar{x} = \frac{\sum fx}{\sum x}$$

where,

$\bar{x}$: Mean

Σfx: Sum of all the values obtained after multiplication of mid values and frequency

Σf: Sum of frequency.

e.g. Calculate mean of following data

No. of Drains Open in Rivers	Mid Value (x)	No. of Rivers (f)	fx
10-20	15	6	90
20-30	25	8	200
30-40	35	10	350
40-50	45	5	225
		Σf = 29	Σfx = 865

$$\bar{x} = \frac{\sum fx}{\sum f} = \frac{865}{29} = 29.82$$

$$\bar{x} = 29.82$$

Short Cut Method

This method is applied for ungrouped data when the frequencies and the values of variables are quite large.

(i) For Individual Observation

e.g. Find the mean marks of 10 students in a class

89, 86, 85, 95, 98, 96, 82, 98, 89, 84

Solution

Let us take assumed mean as a = 90

x	*d = (x-a)*
89	-1
86	-4
85	-5
95	+5
98	+8
96	+6
82	-8
98	+8
89	-1
84	-6
902	Σd = +2

$$x = a + \frac{\Sigma d}{N}$$

$$x = 90 + \frac{+2}{10}$$

$$x = \frac{900+2}{10} = \frac{902}{10} = 90.2$$

x = 90.2

(ii) When Frequencies are Given

Height (cm) *x*	*Number* *f*	*d = x – a*	*fd*
10	5	-20	-100
20	7	-10	-70
30	8	0	0
40	9	+10	+90
50	10	+20	+200
60	6	+30	+180
	Σf = 45		Σfd = 300

Assumed mean =30

$$\bar{x} = a + \frac{\Sigma fd}{\Sigma f}$$

$$\bar{x} = 30 + \frac{300}{45} = \frac{1350 + 300}{45} = \frac{1650}{45}$$

$$\bar{x} = 36.66$$

(*iii*) When Class Intervals are Given

No. of Parasites	No. of Humans (f)	Mid Values (m)	fm	Assumed mean (a)	d = m – a	fd
10-20	10	15	150	35	-20	-200
20-30	20	25	500	35	-10	-200
30-40	30	35	1050	35	0	0
40-50	15	45	675	35	+10	+150
50-60	05	55	275	35	+20	+100
60-70	02	65	130	35	+30	+ 60
	Σf = 82		2780			-90

$$\bar{x} = a + \frac{\Sigma fd}{\Sigma f}$$

$$\bar{x} = 35 + \frac{-90}{80}$$

$$\bar{x} = 33.90$$

(ii) Median

The value which divides the set of variables in to two numerically equal groups after arranging the variables in ascending or descending order is called as median.

Definition

Median is defined as "The middle-most or the central value of the variable in a set of observations, when the observations are arranged either in ascending or in descending order of their magnitude".

Or

According to Gravetter and Wallnav (1989)

"The median is the 50th percentile. Exactly 50% of the individual in a distribution have scales at or below the median ".

Median is calculated by the following formula:

$$Median\ (M) = \frac{N+1}{2}$$

There are several cases to find median *e.g.*

(A) When Data is Grouped

(*i*) When the Case is of Odd Number

e.g. Calculate the median of follwing pH values

4, 9, 6, 2, 12, 5, 8

Solution

Arrange the values in ascending order

2, 4, 5, 6, 8, 9, 12

$$Median\ (M) = \frac{N+1}{2}$$

$$M = \frac{7+1}{2} = \frac{8}{2} = 4^{th}\ Value$$

i.e. Median (M) = 6

(*ii*) When the Case is of Even Number

e.g. Calculate the median of following pH Values

8, 3, 4, 5, 3, 7

Solution

Arrange the values in ascending or descending order

8, 7, 5, 4, 3, 3

$$Median\ (M) = \frac{N+1}{2}$$

$$M = \frac{6+1}{2} = \frac{7}{2} = 3.5$$

It means that median lies between 4th and 5th values

$$\therefore\ M = \frac{3^{rd}\ Value + 4^{th}\ Value}{2}$$

$$M = \frac{5+4}{2} = \frac{9}{2} = 4.5$$

M = 4.5

(B) When Data is Ungrouped

(i) For Discrete Series

In such case arrange values in ascending or descending order and find cumulative frequency. Calculate median by applying the given formula:

$$Median\ (M) = \frac{N+1}{2}$$

Here N = Sum of frequency

e.g. Calculate median of follpwing data

DO Value	7	8	9	10	6	11
Number of rivers	5	4	3	6	2	2

Solution:

DO Value	*Number of Rivers (f)*	*Cumulative Frequency*
6	2	2
7	5	7
8	4	11
9	3	14
10	6	20
11	2	22
	Σf *or* $N = 42$	

$$Median\ (M) = \frac{N+1}{2} = \frac{22+1}{2} = \frac{23}{2} = 11.5 \qquad M = 11.5$$

(ii) For Continuous Series

When class intervals are given apply the given formula to calculate median

$$M = L + \left[\frac{\frac{N}{2} - cf}{f}\right] \times i$$

where,

L: Lower limit value in which median lies

N: Sum of frequencies

cf: Cumulative frequency of the class preceeding the median class.

f: Frequency of median class

i: Class interval of median class.

Median class is a class in which $N/2^{th}$ item lies.

e.g. Calculate median from following table:

Number of Plankton	*Number of Sample (f)*	*Cumulative Frequency (cf)*
6-8	4	4
8-10	6	10
10-12	7	17
12-14	10	27
14-16	5	32
16-18	2	34
18-20	8	42
	Σf *or* $N = 42$	

Here N = 42

$$\frac{42}{2} = 21$$

Here Median lies in the class 12-14

$$M = L + \left[\frac{\frac{N}{2} - cf}{f}\right] \times i$$

$$M = 12 + \left[\frac{21 - 17}{10}\right] \times 2$$

$$M = 12 + \frac{4}{\cancel{10}_{5}} \times \cancel{2}$$

$$M = 12 + 0.8$$

$$M = 12.08$$

(iii) Mode

The variable with maximum frequency is termed as mode.

Definition

According to Arora and Malhan (2001)

"Mode is defined as that value in a series which occurs most frequently. In a frequency distrubution mode is that variable which has the maximum frequency.

According to Gravetter and Wallnav (1987)

"In a frequency distribution, the mode is the score that has the greatest frequency".

Sometimes, the variables have two or more values that have equal frequency in such case the mode is said to be ill-defined and series is called as bimodal or multimodal.

(A) Simple Series

In Simple series the value which is repeated maximum number of times is the mode. *e.g.*

8, 9, 7, 8, 9, 8, 6, 11, 8

Here, Mode is 8 because it is repeated 4 times in a series.

(B) Discrete Series

e.g. Calculate mode from following table:

Number of Plankton	5	6	8	9	11
Number of Samples	5	8	10	16	6

Here, Mode is 9 because it has maximum frequency *i.e.* 16.

When the difference between maximum frequency and the frequency preceding or succeeding it is very small, calculate mode by following technique:

e.g. Calculate mode of following data:

Length of fish (cm)	10	11	12	13	14	15	16	17
Number of fish	2	6	4	8	3	6	5	4

Solution

Length (cm)	*Number (a)*	*Sum of Two (b)*	*Sum of Two Leaving Fish (c)*	*Sum of Three (d)*	*Sum of Three Leaving Fish (e)*	*Sum of Three Leaving First Two (f)*
10	2					
11	6	8				
12	4		10	12		
13	8	12			18	
14	3		11			15
15	6	9		17		
16	5		11		14	
17	4	9				15

Analysing table

Column Number	*Maximum Frequency*
a	13
b	13
c	14,16
d	15
e	13
f	14,17

Since 13 occurs maximum number of times here 13 is mode.

(C) Continuous Series

For continuous of data apply the given formula to obtain mode:

$$Mode = L + \frac{X_1}{X_1 + X_2} \times i$$

where,

L: Lower limit of modal class

X_1: Frequency of class preceding modal class

X_2: Frequency of class succeeding modal class

i: Width of modal class

e.g. Calculate mode by following table:

Number of Zooplankton	*Number of Samples*
0-5	5
5-10	6
10-15	8
15-20	4
20-25	5

Here modal class is 10-15 because maximum frequency is 8

$\therefore$ L = 10

$X_1 = 6$

$X_2 = 4$

i = 5

$$Mode = L + \frac{X_1}{X_1 + X_2} \times i$$

$$Mode = 10 + \frac{6}{6+4} \times 5 = 10 + \frac{6}{10} \times 5$$

Mode = 13

(e) Standard Deviation

Standard deviation also called as root mean square or mean error is universally used. It is used to measure the absolute dispersion.

Definition

It is defined as the square root of the arithmetic mean of the squared deviation of the observations from their arithmetic mean.

Standard deviation is generally denoted by Greek alphabet σ or S. It should be noted that a small standard deviation means a high degree of uniformity of the measurements as well as the homogeneity of the series and vice versa. Thus standard deviation helps in judging the representativeness of the mean.

$$\sigma = \sqrt{\frac{\sum(X-\overline{X})^2}{n}}$$

where,

x: Value of observation

$\overline{x}$: Arithmetic mean

n: Total number of observations

e.g. Calculate the standard deviation of given values

5, 10, 15, 20, 20

$$\text{Arithmetic mean} = \frac{5+10+15+20+20}{5} = \frac{70}{5} = 14$$

X	$X-\overline{X}\,(X-14)$	$(X-\overline{X})^2$
5	-9	81
10	-4	16
15	1	1
20	6	36
20	6	36

$$\Sigma(X-\overline{X})^2 = 170$$

$$\sigma = \sqrt{\frac{\sum(X-\overline{X})^2}{n}} = \sqrt{\frac{170}{5}}$$

$$\sigma = \sqrt{54}$$

$$\sigma = 7.2$$

Sometimes when the mean of the data comes to be a fraction calculate standard deviation by applying following formula/method:

$$\sigma = \sqrt{\frac{\sum(X-\overline{X})^2}{n}} = \left(\frac{\sum X-A}{n}\right)^2$$

where,

x: Value of observation

A: Observed mean

n: Total number of observations

e.g. Calculate the standard deviation of given values

5, 10, 15, 20, 25, 30, 35, 40

Calculation

Let the assumed mean A be 20

X	*X–A*	*(X–A)²*
5	–15	225
10	–10	100
15	–5	25
20	0	0
25	5	25
30	10	100
35	15	225
40	20	400
N = 8	ΣX–A = 20	Σ (X–A)² = 1100

$$\sigma = \sqrt{\frac{\sum (X-\overline{X})^2}{n} - \left[\left(\frac{\sum X-A}{n}\right)^2\right]}$$

$$\sigma = \sqrt{\frac{1100}{8} - \left(\frac{20}{8}\right)^2}$$

$$\sigma = \sqrt{137.5 - 6.25}$$

$$\sigma = \sqrt{130.25}$$

$$\sigma = 11.46$$

When the frequency of data is given calculate standard deviation by following method:

$$\sigma = \sqrt{\frac{\sum fd^2}{n} - \left(\frac{\sum fd}{n}\right)^2}$$

where,

d: X – A

e.g. Calculate mode of following data:

Length of fish (cm)	5	10	15	20	25
Frequency/No.	2	5	7	8	4

Length of Fish (X)	*Frequency (f)*	*d = X – A A = 15*	*fd*	*fd²*
5	2	–10	–20	200
10	5	–5	–25	125
15	7	0	0	0
20	8	5	40	150
25	4	10	40	400
	n = Σf = 25		*Σ fd = 35*	*Σ fd² = 875*

$$\sigma = \sqrt{\frac{\sum fd^2}{n} - \left(\frac{\sum fd}{n}\right)^2}$$

$$\sigma = \sqrt{\frac{875}{26} - \left(\frac{35}{26}\right)^2}$$

$$\sigma = \sqrt{33.65 - 1.81}$$

$$\sigma = \sqrt{31.84}$$

$$\sigma = 5.46$$

(f) Correlation

Correlation is a measure of the relation between two or more variables. The measurement scales used should be at least interval scales, which means that the scale of measurement allows not only to ranking the order items, but also quantify and comparing the sizes of differences between them. Using non-technical language, one can say that the correlation coefficient determines the extent to which value of two variable are proportional to each other.

Most empirical research belongs clearly to one of these two general categories: Firstly in correlation research we do not influence any variable but only measure them and look for relations (correlations) between same set of variables, such as DO and BOD value. Secondly in experiment we manipulate same variables and then measure the effects of this manipulation on other variable: For example a researcher might artificially increase the nutrition level of the culture and then recorded the growth rate at that particular nutrition concentration. Data analysis in experiment also comes down to calculating Correlations between variables, specifically, those manipulated and those affected by the manipulation. However, experimental data may potentially provide qualitatively better information and can demonstrate casual relations between variable. For example if we found that whenever we change variable A then variable B changes, then we can conclude that A influences B.

Correlation Coefficients can range from –1.00 to +1.00. The value of –1.00 represents a perfect negative correlation while a value of +1.00 represents a perfect positive correlation. A value of 0.00 represents a lack of correlation.

Negative Correlation

This relationship shows inverse relation between two variables, which means if one variable increases the other decreases. *e.g*. The relationship between DO and BOD.

Positive Correlation

This relationship between two variables is such that as one variable's value tends to increase, the other variable values also increases. *e.g* Correlation between ammonical nitrogen and Planktonic growth.

The most widely used type of correlation coefficient is Pearson (r), also called linear or Product-moment correlation. It is the basic type of correlation.

Pearson Correlation

Assumes that the two variables are measured on at least interval scales, and it determines the extent to which value of the two variables are proportional to each other. The value of correlation (*i.e.*, correlation coefficient) does not depend on the specific measurement units used; for example, the correlation between temperature (°C) and phytoplankton population (Cells/L). A potential source of problems with the linear (Pearson r) correlation is the shape of the relation. As mentioned before, Pearson measures a relation between two variables only to the extent to which it is linear and deviations from linearity will increase the total sum of squared distances from the regression line even if they represent a true and very close relationship between two variables.

Proportional means linearly related, that is, the correlation is high if a straight line can summarize it. The line is called the regression line or least squares line because it is determined such that the sum of the squared distance of all the data points from the line is the lowest possible. The compact or squared distance will have important function consequences that how the value of the correlation coefficient react to various specific arrangements of data. As we know, the correlation coefficient (r) represents the linear relationship between two variables. If the correlation coefficient is squared, then the resulting value (r^2 or the coefficient of determination) will represent the proportion of common variation in the two variables that is the strength or magnitude of the relationship. In order to evaluate the correlation between variables, it is important to know this magnitude or strength as well as the significance of the correlation. The significance test is for evaluating the significance of difference between correlation coefficient of two samples.

Coefficient of determination is the square of the product-moment correlation or Pearson r between two variable (r^2). It expresses the amount of common variation between the two variables.

$$\text{Karl Pearson's coefficient of correlation (or r)} = \frac{\sum (X_i - \overline{X})(Y_i - \overline{Y})}{n.\sigma_x.\sigma_y}$$

where,

X_i: i^{th} value of X variable

$\overline{X}$: Mean of X

Y_i: i^{th} value of Y variable

$\overline{Y}$: Mean of Y

n: Number of pairs of observation of X and Y

σ_x: Standard deviation of X

σ_y: Standard deviation of Y

In case we use assumed means (A_x and A_y for variables X and Y respectively) in place of true means, then Karl Pearson's formula is reduced to:

$$r = \frac{n\sum xy - \sum x \sum y}{\sqrt{n\sum x^2 - \left(\sum x\right)^2}\sqrt{n\sum y^2 - \left(\sum y\right)^2}}$$

The following steps calculate the coefficient of correlation:

1. Denote one series by x and the other series by y.
2. Calculate $\overline{x}$ and y of the x and y series respectively.
3. Take the deviations of the observations in x series from $\overline{x}$ and write it under the column headed by $dx = x - \overline{x}$. Take the deviations of the observations in y series from $\overline{y}$ and write it in a column headed by $dy = y - \overline{y}$.
4. Square these deviations and write them under the columns headed by dx^2 and dy^2.
5. Multiply the respective dx and dy write it; under the column headed by $dxdy$.
6. Apply the following formula to calculate r and r_{xy}, the coefficient or correlation.

$$r = \frac{\sum dxdy}{\sqrt{\sum dx^2 \times \sum dy^2}}, \text{ or } r = \frac{\sum dxdy}{n\sigma_x \times \sigma_y},$$

Where n is the number of observations in x or y series.

Charles Spearman's Coefficient

Correlation (or rank correlation) is the technique of determining the degree of correlation between two variables in case of ordinal data where ranks are given to the different values of t variables. The main objective of this coefficient is to determine the extent to which the two sets of ranking are similar or dissimilar. This coefficient is determined as under:

$$\text{Spearman's coefficient of correlation (or } r_s) = 1 - \left[\frac{6\sum d_i^2}{n\left(n^2 - 1\right)} \right]$$

where,

d_i: Difference between ranks of ith pair of the two variables;

n: Number of pairs of observation.

As rank correlation is a non-parametric technique for measuring relationship between paired observations of two variables when data are in the ranked form.

Simple and Multiple Correlation Regression Analysis

Regression is the determination of a statistical relationship between two or more variables. In simple regression, we have only two variables; one variable (defined as independent) is the case of the behavior of another one (defined as dependent variable). Regression can only interpret what exists physically i.e., there must be a physical way in which independent variable X can affect dependent variable Y. The basic relationship between X and Y is given by:

$$\hat{Y} = a + bX$$

Where, the symbol $\hat{Y}$ denotes the estimated value of Y for a given value of X. This equation is known as the regression equation of Y on X (also represents the regression line of Y on X when drawn on a graph) which means that each unit change in X produces a change of b in Y, which is positive for direct and negative for inverse relationships. When there are two or more than two independent variables, the analysis concerning relationship is known as multiple correlation and the equation describing such relationship as the multiple regression equation. We here explain multiple correlation and regression taking only two independent variables and one dependent variable (Convenient computer programs exist for dealing with a great number of variables). In this situation the results are interpreted as shown below:

Multiple regression equation can be written as

$$\hat{Y} = a = b_1 X_1 = b_2 X_2$$

Where, X_1 and X_2 are two independent variables and Y being the dependent variable, and a, b_1 and b_2 are constants.

Partial Correlation

Partial correlation measures separately the relationship between two variables in such a way that the effects of other related variables are eliminated. In other words, in partial correlation analysis, we aim at measuring the relation between a dependent variable and a particular independent variable by holding all other variables constant. Thus, partial coefficient of correlation measures the effect of its independent variable on the dependent variable. To obtain it, it is necessary to compute the simple coefficients of correlation between each set of pairs of variables as stated earlier. In the case of two independent variables, we shall have two partial correlation coefficients denoted $r_{xy_1.x_2}$ and $r_{yx_2.x_1}$ which are worked out as under:

$$r_{yx_1x_2} = \frac{R^2_{y.x_1x_2} - r^2_{yx_2}}{1 - r^2_{yx_2}}$$

This measures the effect of X_1 on Y, or more precisely, that proportion of the variation of Y not explained by X_2, which is explained by X_1. Also,

$$r_{yx_2x_1} = \frac{R^2_{y.x_1x_2} - r^2_{yx_1}}{1 - r^2_{yx_1}}$$

in which X_1 and X_2 are simply interchanged, given the added effect of X_2 on Y.

(g) Chi-Square test

It is used to compare the results obtained experimentally with those that obtained theoretically based on hypothesis.

Or

This test is applied to those problems where we study whether the frequency of occurrence of a given event is significantly different from the expected one as theoretically.

Chi-Square test helps us to find degree of discrepancy between observed and theoretical frequency, and determine whether the discrepancy so obtained is due to sampling error or due to chance. Chi-Square is not a parameter but is a non-parametric test as it is computed on the basis of frequencies in a sample and the value so obtained is a statistic.

Karl Pearson first used Chi-square test in1900 for testing statistical hypothesis.

$$\Psi^2 = \sum_{i-1}^{n} \left[\frac{(O_i - E_i)^2}{E_i} \right]$$

O_i: Observed frequency of ith event.

E_i: Expected frequency of ith event.

Properties of Chi-Square test

1. Value of Ψ^2 is always positive.
2. Ψ^2 value can lies anywhere from 0 to ∞.
3. Significance test on Ψ^2 is based on one tailed test, always and of the right side of standard normal curve as Ψ^2 is always positive.
4. Ψ^2 is a statistic therefore it is free from assumption about the form of original distribution from which the observation drawn.

Degree of Freedom

Number of data given in the form of a series of variables in a row or column or number of frequencies put in cells in contingency table, which can be calculated independently is called degree of freedom (ν).

1. $\nu = n - 1$, n is number of variables in the series in a row or column.
2. $\nu = (R - 1)(C - 1)$, R is number of row and C is number of column, when number of frequencies are placed in cells in a contingency table.

Chi-Square Distribution

It is continuous distribution and its probability density function is given by:

$$p\left(\Psi^2\right) = y_0\left(\Psi^2\right)^{1/2(\nu-2)} e^{-1/2\Psi^2}$$

y_0: Constant depends up on the degree of freedom.

Properties of Chi-Square Distribution

1. Always positively skewed.
2. Value increase with the increase in degree of freedom.
3. Mean is the number of degree of freedom.
4. Standard deviation of Ψ^2 distribution $= \sqrt{2\nu}$
5. Different degree of freedom have different shape of curve.

Use of Chi-Square Test

Chi-Square test is used to test the statistical problems.

(*a*) Test of Goodness of Fit

Curved frequency and that of expected frequency are drawn, then the Chi-Square statistic may be used to determine whether the two curves so drawn are fitted good or not. Only one variable is in this test or degree of freedom is $\nu = n - 1$.

(*b*) Test of Independence of Attributes

Chi-Square test is used to show that the principles of classification of attributes are independent and is there is any relationship between attributes. Here attributes are classified into a two way table or contingency table depending on the case. The observed frequency in each cell is known as **Cell frequency**. The total frequency in each row or column of the two way contingency table is known as **Marginal frequency.**

(c) Test of Homogeneity or Test for a Specified Standard Deviation

Chi-Square test is used to test the homogeneity of the attributes in respect of a particular characteristic or to test the population variance. In the case of a specified standard deviation, the test statistic is given by

$$\Psi^2 = (n-1)s^2 / \sigma_o^2$$

where,

s^2: Sample variance and

σ^2: hypothesized value of population variance

Conditions for Using the Chi-Square Test

1. Each of the observations should be independent of each other.
2. The expected frequency of any items or cell should not be less the 10.
3. Total number of observations used in the test must be large *i.e.*, $n \geq 50$.
4. Test cannot be used for estimation of parameter or any value, but can be used only for drawing inferences by testing hypothesis.
5. Totally depend on the degree of freedom.
6. Frequencies used should be absolute, not relative in terms.
7. Observations recorded should be on random basis of sampling.

Working Rule for Chi-Square Test

Chi-Square test is widely used to test the independence of attributes.

1. First of all calculate the expected frequencies on the basis of given hypothesis or on the basis of null hypothesis. Usually in case of a 2 × 2 or any contingency table, the expected frequency for any given cell is worked out as under:

$$\text{Expected frequency} = \left[\frac{(\text{row total for the row of that cell}) \times (\text{column total for the column of that cell})}{(\text{grand total})}\right]$$

2. Obtain the difference between observed and expected frequencies and find out the squares of such differences
3. Divide the quantity $(O_{ij} - E_{ij})^2$ by the corresponding expected frequency to get $\frac{(O_{ij} - E_{ij})^2}{E_{ij}}$ and this should be done for all the cell frequencies or the group frequencies.

4. Find the summation of $\frac{\left(O_{ij}-E_{ij}\right)^2}{E_{ij}}$ value, and this is the Chi-Square (Ψ^2) value.

(h) T-Test

Student's t-Distribution

When the sample is small (i.e., n<10) then the distribution of the standardized variable Z of the statistic t will be far from normality and as a result the normal test cannot be applied. The theory of small (or Exact) sample was developed by Irish statistician William S Cosset who used to write under pseudonym open name of student in 1908. Cosset gave his statistic name as Student's t-distribution or t-test. If $x_1, x_2, \ldots\ldots, x_n$ is a random sample size of n drawn from normal population with mean μ and variance σ^2 (not known) then the student's t statistic (mean) is defined as

$$t = \frac{\frac{\bar{x}-\mu}{}}{\text{(S.E. of mean)}} = \frac{\bar{x}-\mu}{\sqrt{s^2/n-1}}$$

within (n-1) degree of freedom,

where, s = standard deviation of the sample.

Purpose, Assumptions

The t-test is the most commonly used method to evaluate the differences in means between two groups. For example, the t-test can be used to test for a difference in test scores between a group of patients who were given a drug and a control group who received a placebo. Theoretically, the t-test can be used even if the sample sizes are very small (*e.g.*, as small as 10; some researchers claim that even smaller are possible), as long as the variables are normally distributed within each group and the variation of scores in the two groups is not reliably different. *As mentioned before*, the normality assumption can be evaluated by looking at the distribution of the data (via histograms) or by performing a normality test. The equality of variances assumption can be verified with the F test (which is included in the t-test output). If these conditions are not met, then we can evaluate the differences in means between two groups using one of the nonparametric alternatives to the t-test

Application

t-test has following important applications in testing for small sample (n<10).

To test the significance of a single mean, when the data variance σ is unknown, to test the significance of difference between two sample means the data variances being equal and unknown, to test the significance of an observed sample correlation coefficient or difference between means of two sample (dependent samples or paired observations).

t-Test for Independent Samples–More Complex Group Comparisons

It often happens in research practice that we need to compare more than two groups, or compare groups created by more than one independent variable while controlling for the separate influence of each of them. In these cases, we need to analyze the data using Analysis of Variance, which can be considered to be a generalization of the t-test. In fact, for two group comparisons, ANOVA will give results identical to a t-test. However, when the design is more complex, ANOVA offers numerous advantages that t-tests cannot provide (even if you run a series of t-tests comparing various cells of the design).

t-Test for Dependent Samples–Within-group Variation

The size of a relation between two variables, such as the one measured by a difference in means between two groups, depends to a large extent on the differentiation of values within the group. Depending on how differentiated the values are in each group, a given "raw difference" in-group means will indicate either a stronger or weaker relationship between the independent (grouping) and dependent variable. For example, if the mean WCC (White Cell Count) was 102 in males and 104 in females, then this difference of "only" 2 points would be extremely important if all values for males fell within a range of 101 to 103, and all scores for females fell within a range of 103 to 105; for example, we would be able to predict WCC pretty well based on gender. However, if the same difference of 2 were obtained from very differentiated scores (*e.g.*, if their range was 0-200), then we would consider the difference entirely negligible. That is to say, reduction of the within-group variation increases the sensitivity of our test.

Purpose

The t-test for dependent samples helps us to take advantage of one specific type of design in which an important source of within-group variation (or so-called, error) can be easily identified and excluded from the analysis. Specifically, if two groups of observations (that are to be compared) are based on the same sample of subjects who were tested twice (e.g., before and after a treatment), then a considerable part of the within-group variation in both groups of scores can be attributed to the initial individual differences between subjects. Note that, in a sense, this fact is not much different than in cases when the two groups are entirely independent , where individual differences also contribute to the error variance; but in the case of independent samples, we cannot do anything about it because we cannot identify (or "subtract") the variation due to individual differences in subjects. However, if the same sample was tested twice, then we can easily identify (or "subtract") this variation. Specifically, instead of treating each group separately, and analyzing raw scores, we can look only at the differences between the two measures (*e.g.*, "pre-test" and "post test") in each subject. By subtracting the first score from the second for each subject and then analyzing only those "pure (paired) differences," we will exclude the entire part of the variation in our data set that results from unequal base levels of individual subjects. This is precisely what is being done in the t-test for dependent samples, and, as compared to the t-test for independent samples, it always produces "better" results (*i.e.*, it is always more sensitive).

However, the t-test for dependent samples analysis is performed only on the paired differences, "ignoring" the raw scores and their potential differentiation. Thus, the size of this particular difference of 1 will be compared not to the differentiation of raw scores but to the differentiation of the individual difference scores, which is relatively small: 0.2 (from 0.9 to 1.1). Compared to that variability, the difference of 1 is extremely large and can yield a highly significant t value.

Computation of t-Test Values

To test the significance of a mean small samples

$$\text{Standard error of mean S.E. }(\overline{x}) = \frac{s}{\sqrt{n-1}}$$

where, s = sample standard and n = size of sample.

To test the significance of difference between two means (small samples). The steps that are taken in testing the significance of two means or difference between two means in the case of large samples should also be taken in testing significance of two mean or difference between two means in the case of small samples (n<30), except in respect of the application of test statistic, which are discussed above.

$$\text{Estimated standard deviation of population}\sigma' = \sqrt{\frac{n_1 s_1^{\,2} + n_2 s_2^2}{n_1 + n \;\; - 2}}$$

where, s_1 = S.D. of sample 1

s_2 = S.D. of sample 2

$$\text{Standard error of difference=S.E.}(\overline{x}_1 - \overline{x}_2) = \sigma' \sqrt{\left(\frac{1}{n_1} - \frac{1}{n_2}\right)}$$

$$\text{t-test=}\frac{\overline{x}_1 - \overline{x}_2}{S.E.(\overline{x}_1 - \overline{x}_2)}$$

where, $\overline{x}_1$ = Mean of Sample I, n_1 = Size of Sample I

$\overline{x}_2$ = Mean of Sample II, n_2 = Size of Sample II

$S.E.(\overline{x}_1 - \overline{x}_2)$ = Standard Error of Difference.

Chapter 9
Maintenance of Laboratory

Today with the increasing water pollution problem proper monitoring of effluents from industries as well as examination of present status of aquatic bodies is necessary to maintain the ecosystem. Lakhs of monitoring laboratories in India are engaged in examining the physico-chemical and biological characteristics of water and wastewater including pollution control boards, Public health departments, research institutions and consultancies. But here the question arises are these laboratories well equipped according to modern techniques and research? This chapter highlights some of the basic guidelines to maintain the laboratory these guidelines are applicable to all types of laboratories whether small or big.

First of all achievement of safe and healthful workplace is the responsibility of organization. The choice of workplace depends upon the type of analysis to be carried out in the laboratory, facilities available level of accuracy desired and availability of funds.

While maintaining the laboratory following points should be kept in mind.

Proper Ventilation

Laboratory should be well ventilated it should ensure that laboratory air is continuously being replaced, so that toxic or odorous air inside laboratory may not increase.

Separate Sterilization Chamber

Separate sterilization chamber should be made for microbiological work.

Staff Training

Training in safety measures should be an integral part of new employees' introducing to the laboratory besides this regular staff training should be given to include safe methods for dealing with the highly hazardous procedures that are commonly encountered by all laboratory personnel, involving:

1. Inhalation risks (*i.e.* aerosol production), such as using loops, streaking agar plates,
2. Pipetting, making smears, opening cultures, taking blood/serum samples, centrifugation
3. Ingestion risks, such as handling specimens, smears and cultures
4. Risks of percutaneous exposures, through the use of syringe and needle techniques
5. Animal handling that may result in bites and scratches
6. Handling of blood and other potentially hazardous pathological materials
7. Decontamination and disposal of infectious material.

For this reason, continuous in-service training in safety measures is essential

Precautions

- The laboratory should be kept neat, clean and free of materials that are not pertinent to the work.
- Work surfaces must be decontaminated after any spill of potentially dangerous material and at the end of the working day.
- All contaminated materials, specimens and cultures must be decontaminated before disposal or cleaning for reuse.
- Packing and transportation must follow applicable national and/or international regulations.
- When windows can be opened, they should be fitted with arthropod-proof screens.
- Only standards and approved methods should be used in laboratory.
- Sufficient amount of light should be present in laboratory.
- First aid facility should be available in laboratory.
- All workers should follow safety rules.
- Training should be provided to workers for proper working.
- Quality of work should be regularly checked.
- Smoking should be prohibited inside laboratory.
- All chemicals should be correctly and clearly labeled.
- Instruments should be used only for its designed purposes.
- Well-cleaned and sterilized glassware should be used for better results.
- All Instruments should be calibrated time to time.
- Hazardous chemical should be kept in separate area.
- Always follow manufacturers manual for new instrument handlings.

- Separate, cabinets and cupboards should be made with slides glass box for storage of chemicals, glassware etc.
- Working table should have washbasins and shelves for reagents.
- Pipetting by mouth must be strictly forbidden.
- Materials must not be placed in the mouth. Labels must not be licked.
- All technical procedures should be performed in a way that minimizes the formation of aerosols and droplets. Whenever there is an increased risk of aerosolization, work should be conducted in a biological safety cabinet.
- The use of hypodermic needles and syringes should be limited. They must not be used as substitutes for pipetting devices.
- All spills, accidents and overt or potential exposures to infectious materials must be reported to the laboratory supervisor. A written record of such accidents and incidents should be maintained.
- A written procedure for the clean up of all spills must be developed and followed.
- Laboratory waste discharging system should be well planned.
- Use sterilized gloves, clothing, footwear, shields, masks etc.
- Do not use mouth suction to pipette chemicals, pipette bulb should be used to avoid accidents.
- Always prepare fresh solutions for analytical work.
- Floors must be cleaned regularly with disinfectant.
- Precautions should be taken while handling hazardous/poisonous chemicals.
- Spilled chemicals should be cleaned immediately.
- Glasswares or utensils used for laboratory purposes should never be used for preparing food.
- Working area should be clean and free from obstructions.

Cleaning and Sterilization of Glasswares

Clean or wash all the glassware's and jerricans with soda and chromic acid and rinse with double distilled water 2-3 times.

Sample Bottles

Always use 300-ml capacity BOD bottles made of borosil for sample collection. Wash these bottles with washing soda and chromic acid and rinse with tap water followed by distilled water; Wrap the neck and stopper with butter paper with the help of rubber band. Sterilize the bottles in an autoclave at 15 lbs. pressure (121°C) for 15-25 minutes.

Pipettes

Wash Pipettes of different volume size, fitted with cotton plug at the upper end; then wrap in butter paper and sterilize in an autoclave at 15-lbs./inch^2 pressure at 121°C for 15-20 minutes.

Petri-dishes

Wash Petri dishes with washing soda and chromic acid then wrap it in butter paper and sterilized in a hot air oven at 160-180°C for 2-3 hours.

Test Tubes

Wash borosil test tubes and then plug with non-absorbent cotton wool, arrange these tubes in test tube racks and sterilize in an autoclave at 15-lbs./inch^2 pressure at 121°C for 15-20 minutes.

Other Glassware's

Wash other glassware such as conical flasks, beakers, measuring cylinders etc. carefully and then sterilize in hot air oven at 160-180°C for 2-3 hours.

Bibliography

APHA-AWWA-WPCF. 1998. Standard Methods for examination of water and waste water, American Public Health Association, Washington D.C.

Arceivala, S.J. 1981. Waste water treatment and disposal. Marcel Dekkar Inc. New York.

Arora, P.N. and Malhan, P.K. 2001. Biostatistics. Himalaya Publishing House, Mumbai, pp 1-447.

Association of British Chemical Manufacturers and Society for Analytical Chemistry. 1958. Recommended Methods for the analysis of trade effluents, Heffer and Sons.

Biswas, Asit, K. 1998. Water resources. Tata McGraw Hill Publishing Company Limited, New Delhi pp 1-737.

Biswas, K.P. 1949. Common fresh and brackish water algal flora of India and Burma: Records of the botanical Survey of India, Vol. XV, Part I.

Bordner, R.B and Winter, J (ed.) 1978. Microbiological Methods for Monitoring the Environment-water and Wastes. Environment Monitoring and Supp. Lab., Office of R&D, U.S.E.P.A., Cincinnati, Otio, U.S.A.

Botkin, Daniel B. and Keller Edward A. 2000. Environmental Science. John Wiley and Sons. Inc. pp. 1-647.

Carpenter, K.E. 1928. Life in inland waters with special reference to animals. Sidwick and Jackson, London, 1-15 : 1-267.

Chapman, H.D. and Pratt, P.F. 1961. Methods of analysis of Soils, Plants and Waters. University of California.

Daniel, W.W. 1974. Biostatistics : A foundation for analysis in the health science. John Wiley and Sons, Inc.

Day, F. 1878. The fishes of India being a natural history of the fishes known to inhabit the seas and fresh water of India, Burma and Ceylon, vol., I & II, William Dawson and Sons Ltd., London, 20-778.

Douglas, B. 1958. The ecology of the attached diatoms and other algae in small strong stream, J. Ecol. 46: 295-322.

Fassett, N.L. 1940. A manual of aquatic plants. McGraw Hill, New York 1-382.

Gaarder, T. and Gram, H.H. 1917. Investigations on the production of plankton in the Oslofjord. Repp. Cons. Explor. Mer. 144:56-60.

Geldreich, E.E. 1975. Hand Book for Evaluating water bacteriological laboratories. EPA-670/9-75.006. U.S.E.P.A. Cincinnati, Ohio, USA.

Golterman, H.L., Clymo, R.S. and Ohnstad, M.A.M. 1978. Methods for physical and chemical analysis of fresh water. IBP Hand Book No.8, Blackwell Scientific Publications.

Grady, C.P.L and Lim, H.C. 1981. Biological Waste Water Treatment. Marcel Dekker Inc. New York.

Gravetler, F.J. and Wallnav, L.B. 1987. Statistics for the Behavioral Sciences. Tata McGraw-Hill Publishing Co. Ltd. New Delhi pp 1-705.

Jackson, M.L. 1958. Soil Chemical Analysis. Prentice-Hall, London.

James, A. and Evison, L. 1979. Biological Indication of water quality, John Wiley and Sons.

Jhingran, V.G. 1975. Fish and Fisheries of India. Hindustan Publishing Corporation (India) : 1-954.

Katz Max. 1971. Toxicity Bioassay techniques using aquatic organisms. pp 763-800. In : (Ciaccio, LI. ed.) Water and Water Pollution Hand Book, vol.2, Marcel Dekker Inc. New York.

Khanna, D.R. 1993. Ecology and Pollution of River Ganga, Ashish Publishing House, Delhi pp 1-241.

Khanna, D.R. and Bhutiani, R. 2004. Water analysis at a glance. ASEA Publication, Rishikesh.

Kudesia, V.P. 1980. Water Pollution. Pragati Prakashan, Meerut, 1-300.

Lackey, J.B. 1938. The Manipulation and counting of river plankton and changes in some organisms due to formalin preservation U.S. Public Health Reports 53 : 2080-2093.

Lehr, J.H., Gass, T.E., Petty John, W.A. and Demorre, J. 1980. Domestic Water Treatment. McGraw-Hill Book Co.

Leith, W. 1973. The analysis of organic pollutants in water and waste waters. Ann Arbor Science Publs.

Linderg, G.V. 1976. Fishes of the World. A Key to familiar and checklist, John Wiley and Sons, New York, Toronto (Translated from Russian to English) 1-545.

Mancy, K.M. (ed.) 1971. Instrumental analysis for water pollution. Ann Arbor

Manivaskam, N. 1996. Physico chemical examination of water, sewage and industrial effluent, Pragati Prakashan Meerut.

Mark, J. Hammer 1977. Water and Waste water technology, S I version. John Wiley and Sons, Inc. New York, Santa Barbara Chichester, Bribance Toronto, 1-504.

Mathur, R.P. 1982. Water and Waste Waster testing. Nemchand and Bros. Publishers, Roorkee, 1-54.

Metcalf and Eddy, 1979. Water Engineering; Treatment Disposal and Reuse. Tata Mc Graw Hill Publications Co. Ltd., New Delhi.

Minas 1985. Minimal National Standard by Central Pollution Control Board, New Delhi.

Moss, Brian 1982. Ecology of fresh waters. Blackwell Scientific Publications Oxford, 1-332.

National Academy of Sciences. 1977. Drinking Water and Health. Federal Register 43(132) : 357-364.

Needham, J.G. and Needham, P.R. 1972. A guide to the study of fresh water Biology. Holden-Day I.N.C San Francisco. Calif. 94(3) : 1-108.

Nemerow, N.L. 1978. Industrial Water Pollution. Addison-Wesley Publishing Company.

Nikolsky, G.V. 1963. The Ecology of Fishes. Academic Press, London, 1-352.

Odum, E.P. 1971. Fundamentals of Ecology. W.B. Sounders Company, Philadelphia.

Palmer, C.M. 1968. Keys to Water quality indicative organisms (South Eastern United States). Dept. of Interior, Federal Water Pollution Control Administration, Ohio, U.S.A.

Pandey, G.N. 1999. Environmental management, Vikas Publishing House Ltd., New Delhi pp 1-398.

Peter, A.K. 1974. Sources and classification of water pollutants. In : (N. Irving Sox, ed.) Industrial Pollution. Van Nostrand Reinhold Company.

Prescott, G.W. 1969. The Algae : A Review, Thomas Nelson and Sons, London.

Rabosky, J.G. and Koraido, D.L. 1973. Gaging and Sampling industrial waster waters chemical Engineering 80(1) : 111-120.

Schwarebel, J. 1970. Methods of Hydrobiology. Pergammon Press.

Shrivastava, G.L. 1986. Fishes of Eastern Uttar Pradesh. Vishwavidyalaya Prakashan, Varanasi, 1-163.

Stern, A.M. and Walker, C.R. 1978. Hazard assessment of toxic substances : Environmental effect testing of organic chemicals and ecological effect testing.

pp. 81-131. In : John Cairns Jr, Dickson, K.L. and Maki, A.W. ed.) Estimating the Hazard of chemical substances in aquatic life. ASTM STP. 657.

Tranter, D.J. (Ed.) 1908. Reviews of Zooplankton Sampling Methods. UNESCO Switzerland.

Traversy, W.J. 1971. Methods for chemical analysis of waters and waste water. Dept. of fisheries and forestry, Ottawa.

Trivedy, R.K. and Goel, P.K. 1986. Chemical and Biological Methods of Water Pollution Studies. Environmental Publications, Karad. pp 1-244.

U.S.E.P.A. 1976. Methods of Chemical Analysis for Soil Survey Samples. NZ Dept. Sci. and Ind. Res. Soil biol. Bull 12.

Uberoi, N.K. 2004. Environmental Management, IInd edition Excel books, New Delhi, pp 1-414.

Vollenweider, R.A. (Ed.) 1969. A manual of methods for measuring primary production in aquatic Environment. IBP handbook No. 12 Black well scientific publications.

Vollenweider, R.A. (Ed.) 1974. A manual of methods for measuring primary production in aquatic environment. IBP hand Book No.12 (ed. II) Blackwell Scientific publications.

Ward, H.B. and Whipple, G.C. 1992. Fresh Water Biology IInd Edition, Edited by W.T. Edmondson, International Books and Periodicals Supply Service, New Delhi.

Welch, P.S. 1948. Limnological methods. The Blackiston Co., Philadelphen 1-381.

Wilson, A.L. 1974. The chemical analysis of water. Society for Analytical Chemistry, London.

WPCF. 1976. Operation of waste water treatment plants. A manual of practice, Map/ 11, water pollution control federation, Washington, D.C.

Index